ISBN 978-1-332-02284-7
PIBN 10270488

1 MONTH OF
FREE
READING

at

www.ForgottenBooks.com

By purchasing this book you are eligible for one month membership to ForgottenBooks.com, giving you unlimited access to our entire collection of over 700,000 titles via our web site and mobile apps.

To claim your free month visit:

www.forgottenbooks.com/free270488

English
Français
Deutsche
Italiano
Español
Português

www.forgottenbooks.com

Mythology Photography **Fiction**
Fishing Christianity **Art** Cooking
Essays Buddhism Freemasonry
Medicine **Biology** Music **Ancient
Egypt** Evolution Carpentry Physics
Dance Geology **Mathematics** Fitness
Shakespeare **Folklore** Yoga Marketing
Confidence Immortality Biographies
Poetry **Psychology** Witchcraft
Electronics Chemistry History **Law**
Accounting **Philosophy** Anthropology
Alchemy Drama Quantum Mechanics
Atheism Sexual Health **Ancient History**
Entrepreneurship Languages Sport
Paleontology Needlework Islam
Metaphysics Investment Archaeology
Parenting Statistics Criminology
Motivational

THE MICROSCOPE

IN THE

BREWERY AND MALT-HOUSE

CHAS. GEO. MATTHEWS, F.C.S., F.I.C., Etc.

AND

FRANCIS EDW. LOTT, F.I.C., A.R.S.M., Etc.

Illustrated by Steel Engravings, Woodcuts, Chromo-Lithographs

NEW YORK
D. APPLETON AND COMPANY
1, 3 AND 5 BOND STREET
1889

THE MICROSCOPE

IN THE

BREWERY AND MALT-HOUSE

BY

CHAS. GEO. MATTHEWS, F.C.S., F.I.C., ETC.

AND

FRANCIS EDW. LOTT, F.I.C., A.R.S.M., ETC.

*Illustrated by Steel Engravings, Woodcuts, Lithographs, and
Chromo-Lithographs*

NEW YORK
D. APPLETON AND COMPANY
1, 3, AND 5 BOND STREET
1889

5728
२३/९/२०

To

Mons. Louis Pasteur

THIS BOOK IS INSCRIBED BY THE AUTHORS, IN GRATEFUL
APPRECIATION OF THE HIGH SCIENTIFIC AND PRACTICAL VALUE
OF THE WELL-KNOWN RESEARCHES IN CONNECTION WITH
FERMENTATION UNDERTAKEN BY HIM IN YEARS PAST; AND IN
ADMIRATION OF THE GENIUS DISPLAYED IN THIS AND OTHER
BRANCHES OF SCIENTIFIC INVESTIGATION.

September, 1889.

TABLE OF CONTENTS.

CHAPTER I.

DESCRIPTION OF PLATES.

Compound Monocular Microscope (*Frontispiece*).

PLATE I. (*to face page* 24).

Comparison of Micrometer lines with Metric and Inch Scales.

PLATE II. (*to face page* 34).

Fig. 1.—Development of Yeast.
„   2.—Sporulation of Yeast (after Reess).

PLATE III. (*to face page* 42).

Fig. 1.—Burton Yeast.
„   2.—London Yeast.

PLATE IV. (*to face page* 44).

Fig. 1.—Deteriorated Yeast.
„   2.—   Ditto.

PLATE V. (*to face page* 44).

Fig. 1.—Saccharomyces Pastorianus.
„   2.—Caseous Yeast No. 1 (left hand).
        Ditto   No. 2 (right hand).
„   3.—Saccharomyces Ellipsoideus (left hand after Reess).
        Ditto   ditto   (right hand after M. and L.)
„   4.—Saccharomyces Minor.
„   5.—Saccharomyces Exiguus (left hand after Reess).
        Ditto   (right hand after M. and L.)
„   6.—Mycoderma Vini (left hand, Aerobian form).
        Ditto   (right hand, Submerged form).

Fig. 8.—Bacillus Subtilis.
„   9.—Bacillus Ulna.
„ 10.—Bacillus Leptothrix.
„ 11.—Spirillum Tenue.
„ 12.—Spirillum Undula.

PLATE XI. (*to face page* 128).

The Forcing Tray in working order (from a photograph).

PLATE XII. (*to face page* 136).

Forced Beer Sediments.

Fig. 1.— Normal residue of S. Cerevisiæ.
„   2.—Residue with S. Pastorianus.
„   3.·   „   „   Caseous ferment, No. 1.
„   4.—   „   „   B. Subtilis, etc.
„   5.—   „   „   Sarcina, etc.
„   6.—   „   swarming with B. Lactis and B. Subtilis.

PLATE XIII. (*to face page* 140).

Fig. 1.—Palea, both layers.
„   2.—   „   fibres of outer layer.
„   3.—   „   outer layer.
„   4.—   „   fibres of inner layer.

PLATE XIV. (*to face page* 142).

Fig. 1.—Pericarp.
„   2.—Testa.
„   3.—Diagram Section of Barley-corn (after Holzner).

PLATE XV. (*to face page* 142).

Longitudinal Section of Barley-corn (reduced from Lintner).

PLATE XVI. (*to face page* 142).

Fig. 1.—Transverse Section of Barley-corn.
„   2.—   Ditto, some days after germination has proceeded
(from a photograph).

DESCRIPTION OF WOODCUTS.

1A

LITERATURE CONSULTED DURING THE PREPARATION OF THIS WORK.

BOOKS.

" How to work with the Microscope." Lionel S. Beale.

" The Microscope." Dr. Carpenter.

" The Microscope in Theory and Practice." Nägeli and Schwendener.

" The Student's Handbook to the Microscope." A Quekett Club man.

" Preparing and Mounting Microscopic Objects." Thomas Davies.

" Etudes sur le Vin." L. Pasteur.

" Etudes sur la Bière." L. Pasteur ; and Translation of same entitled " Studies on Fermentation." Faulkner and Robb.

" Fermentation." Schützenberger.

" Microbes, Ferments, and Moulds." Trouessart.

" Botanische Untersuchungen über die Alkoholgährungspilze." Dr. Reess.

" Bacteria and Yeast Fungi." Grove.

Cantor Lectures on " Yeast." A. Gordon Salamon.

" Die Spaltpilze." Dr. W. Zopf.

" Lectures on Bacteria." De Bary ; translated by Garnsey and Balfour.

" Practical Bacteriology." Crookshank.

" Fungi." Cooke and Berkeley.

" Microscopic Fungi." M. C. Cooke.

"Elementary Biology." Prof. Huxley and H. N. Martin.

"Nachrichten über den Verein Versuchs und Lehranstalt für Brauerei in Berlin. Die Sarcina-Organismen der Gährungs-Gewerbe." Dr. Paul Lindner.

"Die Micro-organismen der Gährungsindustrie." Alfred Jorgensen; and Translation of same, by Dr. G. H. Morris.

"Malzbereitung und Bierfabrikation." J. E. Thausing.

"Handbuch der Spiritusfabrikation." Dr. Maercker.

"Lehrbuch der Bierbrauerei." Dr. Carl Lintner.

"Untersuchungen aus der Praxis der Gährungsindustrie. Emil Chr. Hansen.

"Infusoria." A. Pritchard

Transactions of The Laboratory Club, Vol. I.

Transactions of The Burton-on-Trent Natural History Society, Vol. I.

REPORTS, JOURNALS, ARTICLES, &c.

Reports of the Carlsberg Laboratory, 1878 to 1888.

Journal of the Royal Microscopical Society, 1887-8.

Journal of the Quekett Microscopical Club.

Journal of the Chemical Society.

Journal of the Society of Chemical Industry, 1882 to 1889.

"Brewing Trade Review," 1887 to 1889.

"Brewer's Journal," 1880 to 1889.

"Brewer's Guardian," 1880 to 1889.

Articles by Dr. Maddox, 1886 to 1889, in Diary for the Brewing Room, A Boake.

INTRODUCTORY PREFACE.

IN these days when there are few Breweries which do not possess a Microscope, it would seem desirable that the instrument should not fall a prey to the casual or uninstructed observer, but should rather, by the knowledge and skill of those that use it, be made a means of controlling the processes of Malting and Brewing. A Brewer in becoming practically acquainted with the Microscope as a controlling agent in his process, raises, to use a figure of speech, a part of the line of fortification which science provides against the hurtful or injurious influences declaring themselves, when Brewing operations are not conducted with the intelligence and skill that they ever increasingly require.

The production of a special treatise on the microscope as applied to Brewing, was first contemplated by the authors during the delivery of a course of lectures on this subject to some young brewers. As the lectures were re-delivered to successive groups of students, the impression already gained, namely,—that a real requirement existed amongst Brewers for precise and condensed instruction in the handling of the microscope,—became a very strong one indeed, and the present work was undertaken. As the writing advanced it was deemed desirable, in order to make the work as complete as possible, to cover more ground than the occasion at first seemed to demand. The original lectures, however, constitute the nucleus of the work.

The chief aim of the authors has been to collect within a convenient space and without undue elaboration, matter that appears to them to be of undoubted value in its application to Brewing and Malting; and they believe that a good deal of information has been incorporated at the same time, that has not hitherto been adequately dealt with in print. The fact that a large part of the information is drawn from works of undisputed excellence is fully recognized by the authors; but the works are many—as may be judged from the list of authorities quoted—and the expense involved in their purchase would be very considerable, besides which, the search amongst authorities for required information involves the expenditure of no little time and trouble.

With these explanations, and trusting that their efforts have been attended by a reasonable amount of success, the authors hopefully tender this treatise to the judgment of those who are interested in the Industries with which it seeks to identify itself.

The authors would here express their cordial thanks to the friends in Burton (especially the members of the Chemical Club), and elsewhere, who have materially aided them by useful suggestions, loans of photographs, assistance in corrections of MS., revision of proofs, etc. Where such general kindness has been experienced it seems invidious to make any distinction by name.

The authors also wish to record their thanks to Mr. J. E. Wright for the great care and attention bestowed on the drawings bearing his name.

Bridge Chambers,
 Burton-on-Trent.

CHAPTER I.

THE MECHANICAL ARRANGEMENTS OF THE MICROSCOPE.

BEFORE proceeding to discuss the various uses to which the microscope may be applied by the brewer and maltster, it is essential that a fair understanding of the mechanical construction of an ordinary instrument should be arrived at. Knowledge of the optical principles on which the action of the lenses depends, though a desirable acquisition, must from our point of view be looked upon as a matter of separate study ; and it will therefore be necessary to touch only in the briefest manner on a few purely optical considerations. We will then in this first chapter give a general, followed by a more special, description of the parts of what is known as the Compound microscope.*

Referring to Fig. 1, the entire frame-work there represented, to which various movable accessories of the microscope may be adapted, constitutes the Stand, consisting of the tube A and the part A' immediately supporting it, called respectively the Body and the Limb ; the Stage or object carrier B, and the Foot C, this last carrying the whole weight of the instrument, and being, when well contrived, adjusted so as to secure a maximum of steadiness.

---

* A single magnifying lens or Simple microscope is of no special use in connection with brewing matters, being used chiefly for the examination and dissection of comparatively large objects under a low magnifying power.

A microscope having a single tube is known as a Mono-cular—one with a double tube as a Binocular—microscope.

Into the tube at the upper end (Fig. 1 a) slides the Eye-

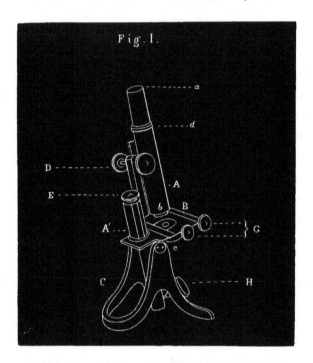

piece (Fig. 2), generally consisting of two lenses with a diaphragm or stop between them. The lens nearest the eye of the observer is called the eye-lens, the other the field-lens, whilst the screw-threaded socket at the other end of the tube (Fig. 1 b) carries the Object glass or Objective (Fig. 3), the most important of the optical parts of the instrument. The screw-thread as a rule is of such a diameter as to admit of objectives by different makers being used with the same tube and stand.*

The body of the microscope is usually controlled by two movements termed Adjustments. Firstly, the larger milled-headed screws (Fig. 1 D) causing the tube by a rack and

---

* That is to say, a standard has been agreed upon so as to render objectives of different microscopes interchangeable, but the makers do not seem to exactly work up to it.

pinion to slide through a vertical distance of some two to four inches. Secondly, the smaller milled-headed screw (E) acting either on the whole tube, or on a socket at its lower end, this last having sometimes an extra play of about three-sixteenths of an inch upon a spring independent of either of the adjustments; this is to protect the objective if it should be impelled by accident against a glass slide or other rigid body, such as the stage itself. The movements being imparted by a fine-threaded screw, may be made as small as desired: the arrangement is used for focussing with high powers, and is known as the Fine adjustment, the one first

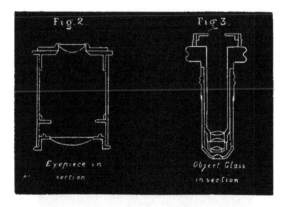

mentioned being termed, in contradistinction, the Coarse adjustment.

The plane surface with a central opening or Stage for carrying slides may be either of metal or glass, with clips, or a ledge to retain the glass slide. It may be provided with Movements, which are ordinarily rectangular; that is, by the use of milled-headed screws (Fig. 1 G) attached to the stage, the slide may be caused to move in directions approaching to or receding from the observer, or from side to side, the two sets of directions being at right angles to each other; or the movements may be compounded into diagonal directions by using both milled-heads simultaneously. A circular movement of the Stage is sometimes provided, but it is not essential to an instrument designed for Brewery purposes; neither, indeed, are the rectangular movements, but they are a great convenience, and are regretfully dispensed with by anyone accustomed to their use. Any receptacle for accessories immediately underlying the Stage is called the Sub-stage (Fig. 1 e). Here, a diaphragm (Fig. 4 aaa), an arrangement to regulate the passage of light to the object under examination is usually found, and is practically indispensable for good definition with high magnifying powers. It consists, generally, of a perforated circular plate, rotating on a centre pin as sketched, the apertures being circles of different diameters; though for

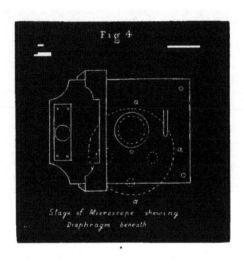

illumination of special objects, other shaped openings are sometimes included. A very elegant form is Collins' "Iris" or graduating diaphragm, in which the aperture may be regulated by a screw, from the smallest circle to a considerable opening. Small perforated discs or "Stops" of different apertures are occasionally made to fix underneath the object instead of the movable diaphragm.

To secure the best defining power of the lenses, especially with high powers, a piece of apparatus called a Stage Condenser (Fig. 5)—which is as a rule, Achromatic—is very useful. It fits into the Sub-stage, and consists of an arrangement of lenses contrived to concentrate light on the object under observation. Where this adjunct is employed, the diaphragm is often placed underneath it as in the figure: it then exercises a first control on the amount of light passing to the object.

Another accessory of the Sub-stage is the Nicol's prism, which constitutes part of the Polarizing apparatus ; a second Nicol's prism—the Analyser—fitting into the tube of the microscope just above the Objective. These accessories

Fig. 5.

Achromatic Condenser
and Diaphragm.

are by no means necessary in a Brewery microscope, but might be of some use for special work.

An indispensable adjunct to the microscope stand is the apparatus for reflecting light on to transparent objects placed on the Stage : for this purpose a double mirror, on a jointed arm, is usually provided, occupying the position indicated by H Fig. 1, having one surface plane and the other concave, the action of which reflectors respectively will be explained later.

In the case of Binocular microscopes, two images are

obtained by a portion of the rays of light from the Objective being diverted by a small prism into the second tube of the instrument, which is usually joined at an angle to what may be called the main tube. An image is thus provided for each eye, and the two eye-pieces are moved simultaneously by a rack and pinion like the coarse adjustment.

With a microscope such as that outlined in Fig. 1, the

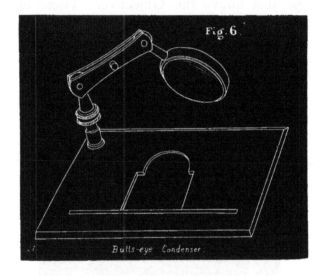

Bulls-eye Condenser.

tube can be lengthened by a sliding piece called the Draw tube, the junction being at *d :* the object of this is to increase the amplification, the effect being similar to that obtained by using a higher power eye-piece.

Amongst necessary appliances is the Bull's-eye condenser, which may be on a separate stand as in Fig 12 A, but is more convenient when it can be attached to the Stage (Fig. 6), and should be provided with a universal movement as indicated. Its use is to concentrate the light on to an opaque or semi-opaque object.

In absence of sunlight it is desirable to have a good source of artificial light to fall back upon ; any of the following may serve :—Firstly, an Argand gas burner on a vertical stand, which is the more convenient if it has a telescopic slide for raising or lowering the burner, and a blue or neutral tint glass cylinder is to be preferred to the ordinary white glass. Secondly, a Paraffin lamp, with blue or neutral tint glass chimney, or a copper chimney is sometimes employed, having an eye or aperture $\frac{1}{2}$ to 1 inch wide, of tinted glass ; or a cylindrical porcelain shade may surround the glass chimney, having a portion cut out to let a certain amount of light issue from the lamp. Amongst more expensive illuminating apparatus, an incandescent electric lamp fitted on a movable arm, is a very neat and effective source of light, and has much to recommend it where the microscope is used intermittently. The new incandescent gas burners of the Clamond and Welsbach pattern yield a very nice steady light.

As regards smaller apparatus. For drawing or sketching with the microscope a Camera Lucida, or Beale's neutral tint reflector, is often employed attached to the eye-piece ; of the two forms the reflector is by far the cheaper, and acts almost as well as the Camera Lucida, which last includes a small glass prism in its structure. The mode of employ-ment of these appliances is described under " Manipulation." Forceps or pincers contrived to fix on the Stage are sometimes useful for holding an object which it is not con-venient to put on a glass slide. A dozen or two of glass slips of the ordinary size, 3 in. by 1 in., and $\frac{1}{2}$ oz. of cover glasses from $\frac{5}{8}$ to $\frac{7}{8}$ in. diameter, may be provided. Cir-cular cover glasses are more conveniently cleansed than squares, as they do not break so easily. With combina-tions not exceeding 300 to 400 diameters, a cover glass of some strength may be employed, as very fragile ones provide a constant source of annoyance by breakage.

There are, of course, innumerable accessories for special kinds of investigation, but the microscope as used by the brewer does not require them. A convenient Stand, with one good eye-piece and two objectives of low and high power respectively, Stage rectangular movements, an Achromatic condenser, a Bull's-eye condenser, and a good artificial source of light (should this be required), constitute pretty well the whole of the apparatus necessary or desirable.

We will now enter into some further detail in elucidation of the action of some of the parts of the instrument already referred to, and thus pave the way to manipulation pure and simple.   The Mirrors, or reflecting apparatus, call for early

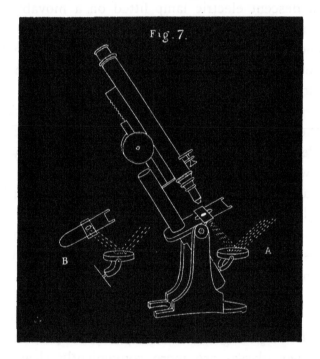

Fig. 7.

consideration, and in connection with Brewing matters the Concave or hollowed mirror is of the greater importance ; this form of reflector concentrates the light to a certain point or "focus" some two or three inches from the centre of the mirror, as shown in Fig. 7 A, and is used in conjunction with high power objectives.   The best position of the mirror may be determined, experimentally, by putting a flat piece of oiled tissue paper or tracing paper on the stage, and moving the mirror vertically till a small disc or spot of light is shown.   The action of the Plane or flat mirror is shown in the small sketch B, appended to Fig. 7.   Here the light, instead of being concentrated, is reflected in parallel rays, and consequently with small objects and object-glasses, a large portion does not impinge upon them at all.   The illumination is, however, quite adequate and satisfactory for objects viewed under low powers of magnification.

We may now deal with the lenses of the microscope as included in the Eyepiece and Objective.   Their action is dependent on the optical principle known as Refraction, or the bending that rays of light undergo when entering a medium of different density, a certain amount of the light being at the same time absorbed or lost.   The degree of refraction is determined by the curvature of the lens and density of glass, high magnifying power being concurrent with great curvature, high refraction, and short focal length or Working Distance ; this last being the interval between the front lens of an objective and the object examined, when the latter is in proper focus.   With high power objectives the object must be very close to the lens, and at a proportionately greater distance as the magnification is less.

Fig. 8 a, b, c shows sections of the Lenses employed in the construction of the microscope, viz., double-convex, plano-convex, and plano-concave, the last-mentioned being used to modify the course of the rays passing through convex lenses to obviate certain imperfections, the nature of which should be understood so as to aid in their detection. One of these imperfections is called Spherical Aberration, and it is rendered obvious by viewing through the microscope a glass slide on which a fine network of squares is ruled.   Fig. 9 B represents what is seen with a proper performance of the properly corrected instrument, whilst the distorted appearance of A and C indicate opposite kinds of aberration, caused by lenses imperfectly corrected. The greater the distortion, the more faulty of course are

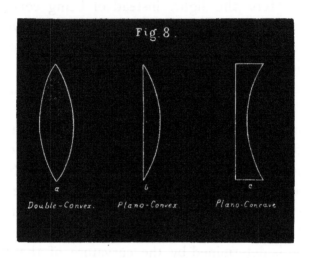

Fig. 8.

Double-Convex.   Plano-Convex.   Plano-Concave.

the lenses. Eye-pieces and objectives thoroughly corrected and free from Spherical aberration are said to be Aplanatic.

Another imperfection of the lenses is that termed Chromatic aberration.   It is the cause of the tinting of

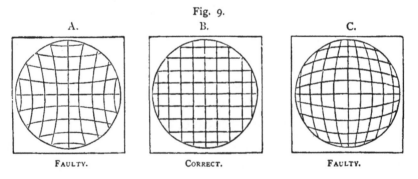

Fig. 9.

A.   B.   C.

FAULTY.   CORRECT.   FAULTY.

colourless objects, and of the coloured fringes so frequently seen surrounding objects viewed through imperfect instruments.   Lenses free from this defect are said to be Achromatic.

The Chromatic and Spherical aberration of a lens may

be diminished by reducing the aperture with a stop or diaphragm, so that only its central portion is employed, but complete correction is only secured by utilizing different shaped lenses, as already indicated, and lenses of different kinds of glass. Objectives of the cheaper kind and especially those of foreign manufacture, have often only a front lens, but the majority of good objectives are built up in the compound form, each lens consisting of two kinds of glass of different optical properties, cemented together with a transparent medium such as Canada Balsam ; and the parting or cracking of the said medium may render an objective practically useless until re-cemented. Fig. 10 shows the arrangement of three pairs of lenses, 1, 2, 3 ; each pair formed of a double convex of crown glass, and a plano-convex of flint glass.

A considerable variety of magnifying power may be obtained by altering the position of lenses in respect to each other and to the object, as shown in the employment of the draw tube ; amplification may be obtained in this way or by using higher power eye-pieces, but in the latter case often at the expense of good definition ; for defects of the object glass which are not perceptible when the image it forms is but moderately enlarged, are brought into prominence when the imperfect image is magnified or amplified to a much greater extent ; so that in practice it is found better to vary the power by employing objectives of different magnification.

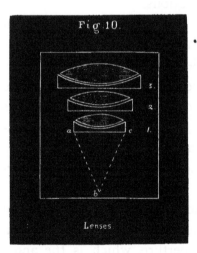

Fig. 10.

Lenses.

Eye-pieces are made of various magnifying powers, but always comparatively low ones ; the range is generally indicated by letters A, B, C, etc., or by numerals, 1, 2, 3, etc., the power increasing from A and 1 respectively.

Object glasses or Objectives are usually designated by their focal distance from the object, viz., 1 in., $\frac{1}{4}$ in., $\frac{1}{8}$ in., and so on, but in nearly all cases the distance at which they focus is less than that implied by the figures, which consequently give an imperfect idea of the real magnifying power. Generally speaking, objectives range from 4 in., giving with an A eye-piece some 10 diameters' magnification, to $\frac{1}{50}$ in. giving 3,000 diameters ; but for Brewers' purposes two objectives, a $1\frac{1}{2}$ in. or 1 in. giving 30 to 50 diameters, and a $\frac{1}{6}$ or $\frac{1}{8}$ in. yielding 300 to 400 diameters—according to the particular maker—suffice. With these objectives, one good eye-piece a little stronger than an ordinary A, should be provided ; or if expense is not so much an object, both A and B eye-pieces may be included. Many opticians now provide tables in their catalogues giving the magnifying power of the combinations.

We may here remark that a really good combination giving only 200 diameters of magnification, will show Yeast and Bacteria with considerable distinctness of detail as regards the former, and of size and shape as regards the latter ; and it is far preferable to work with excellent lenses magnifying some 200 diameters, than with a poor combination magnifying double as much, for the latter case means constant annoyance and irritation from the imperfect performance.

The amount of light admitted by an Object glass is of. considerable importance, and depends in great measure on what is called the Angle of Aperture, which is the angle formed by two lines from opposite sides of the aperture of the Objective with its focus. (See Fig. 10 a, b, c.) Glasses with a high angle of aperture admit much light, but focussing so close to the object they entail considerable inconvenience in general work ; those of medium angle

are preferable, combining as they should, Power of Penetration and Brightness of Field. The latter term speaks
for itself; by the former, is meant the capability of the glass
to give a correct view of an object possessing an appreciable depth. Power of penetration should not be con-
founded with Resolving power, which is the capability of
resolving or dividing the component parts of a minute
object, such as the markings on Diatoms, or the closely-
ruled lines of a test plate: this resolving power is dependent also on angle of aperture, the higher-angled apertures
having a greater resolving power. It will thus be seen
that this quality is opposed to that of penetration, which is
possessed by glasses of low or moderate aperture, and that
the two requisites can only be combined in the same objective by some sacrifice of each. The purpose for which the
instrument is required must govern the choice of " powers."

Another desideratum in an objective is Flatness of Field,
which means that the whole of a large flat object should
be in correct focus at once, even to the extreme margin of
the field of view ; and the same correctness of focus should
be exhibited by objects lying in the same plane. This
quality is of the most importance in the lower powers
with which large objects are usually examined ; in the
case of glasses of short focus, as a $\frac{1}{4}$ in. or higher power,
the object is usually a minute one, and generally placed in
the centre of the field ; and if in the margin of it, the slight
alteration of focus necessary, causes little trouble.

The varying refraction of the thin glass, covering an
object, renders an adjustment of the higher power objectives necessary, and especially so in glasses of high angle
of aperture ; it is usually effected by altering the distance
between the front and second pair of glasses. An engraved
line on the brass mount shows the point to which the lens
should be set for *uncovered* objects. Its adjustment for
covered objects is effected in the following manner : —

Arrange the objective as if for an *uncovered* object. Focus any *covered* object by moving the tube of the microscope ; next move the milled adjustment ring of the objective till particles of dust on the upper surface of the cover-glass are brought into focus.   The objective is now corrected for the thickness of the cover-glass, and it only remains to re-focus the object with the tube adjustments.

Many of the high power objectives now in use are worked on the immersion system, which consists in the interposition of a drop of water or oil—generally Cedar oil—between the front lens of the objective, and either the object itself or its cover-glass.   It is of course, requisite that the objective should be specially corrected for such use.   The advantages gained are a considerable increase of working distance and penetration.

We will conclude this chapter with a few words on the choice of a microscope, first summarizing the qualities of a really good instrument.   They are :—

A fairly large and well-illuminated field of view.

Freedom from Chromatic and Spherical aberration.

Good definition and penetration.

Flatness of field.

Unless the Brewer has had some experience, it is better, in purchasing a microscope, to secure the good offices of someone who knows what a Brewery microscope should be capable of doing, and what is really good value for the amount of money it is purposed to expend ; for with the best intentions on the part of the maker, his want of appreciation of the special purpose to which the instrument is to be applied, may cause him to forward a disappointing or unsuitable article.

There are, at the present time, so many makers of excellent microscopes at a moderate price, such as—Messrs. Baker, Beck, Browning, Crouch, Steward, Swift, and Watson ; and amongst foreign makers, MM. Seibert and Zeiss

—that it would be invidious to make any special selection for recommendation ; suffice it to say that the authors have made the chief part of their observations with the more complete form of Swift's College microscope, than which, at the price, no more satisfactory instrument has ever been in their hands. With Messrs. Swift & Son's permission, a drawing of this microscope is given as the frontispiece.

CHAPTER II.

On Manipulation.

IN the first place it is obviously of necessity that the lenses of the microscope should be scrupulously clean. This is best secured by carefully wiping them with a cleansed and softened wash-leather, glass-cloth, or silk handkerchief; some soft fabric that does not "lint" is essential. Specks of dust on the glasses of the eye-piece may be detected by turning it round whilst looking through the instrument, as any such specks will be found to move with the eye-piece. In cleaning objectives great care must be exercised, and it is seldom necessary to interfere with their inner glasses.

The same attention should be occasionally bestowed on the Stand, and where the microscope is in frequent use, it may conveniently be kept under a Bell-glass, or glass shade, with chenille edging to exclude dust, in which case there is no objection to the powers remaining attached. The instrument must not stand in a damp place, and on no account let any liquid accidentally taken up by the objective, remain and dry upon it. Especial caution must be exercised in this respect with reagents used in the examination of an object. Ordinary care should obviate any contact at all between the objective and substances under examination. Oil or water immersion lenses should be cleaned after use.*

Glass slips and cover-glasses after use, if not immediately cleaned and dried, may be placed in separate vessels containing water; this precludes the nuisance of their becoming cemented together by the drying up of liquids contained between them. Two small jam-pots are sufficiently good receptacles, that for the cover-glasses being the smaller, and having preferably a curved bottom; the water should be renewed frequently, and if slightly acidulated, deposition of Carbonate of Lime is prevented; or distilled water may be employed. After cleaning and drying, it is a good plan to keep the cover-glasses and slips in a wash-leather case, sewn into separate compartments. (Fig. 11.)

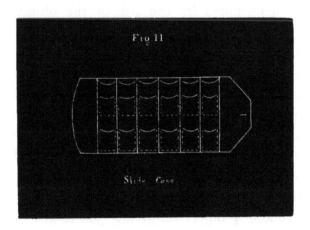

Fig. 11

Slide Case.

The microscope should stand on a steady table or desk of convenient height, say from 24 to 30 inches, according to the size of the instrument. Should the room be subject to vibrations from machinery, etc., it is well to have the legs of the table on thick India-rubber pads, and the microscope on a sheet of the same material. It is decidedly better to work seated, and a revolving study chair is a great convenience. The instrument should be placed in a good

* A little turpentine may be used if necessary to remove Cedar and other oils.

3

light, preferably a N.W. to N.E. aspect, as direct sunlight is unsuitable for the higher magnifying powers : diffused light, such as that from large white clouds, gives the best field. The microscope may be placed fairly close to the window, but care must be taken to have the mirror opposite a clear and clean pane.

It is convenient to have a small sink and water-tap close by, with a shelf for sample bottles and glasses, and a draining rack.

In attaching objectives it is advisable to hold them up with the left hand, whilst screwing on with the right fore-finger and thumb.

The mode of treatment, the particular combination of eye-piece and objective, and the degree of illumination, are necessarily determined by the size of the object, and the condition in which it exists. With objects such as a Hop cone or Barley corn, a magnification of 30 or 40 diameters, —secured by the A eye-piece and $1\frac{1}{2}$ or 1 in. objective— would be adequate for a general examination, the object being simply placed on a glass slide, and illuminated by the Bull's-eye condenser as in Fig. 12. Successive portions of the above-mentioned objects might then be taken, and finally, the smallest portions examined by the high power combinations, such as A eye-piece and $\frac{1}{8}$ objective, or B eye-piece and $\frac{1}{4}$ objective, giving 300 to 400 diameters ; and transmitted light from each mirror tried, as well as the reflected light from the Bull's-eye condenser.

The observer will soon notice that the position of every-thing viewed through the ordinary microscope is inverted, and for a time this will be found a difficulty, especially when working the stage movements : an appliance termed an Erector restores the position, but is seldom used ; practice removing the difficulty.

When the separate particles of a substance are invisible, or barely visible to the unassisted eye, and insoluble in water, it is often advantageous to examine them in this latter medium as well as in the dry state, equal illumination and a flatter field being secured. Finely divided substances that are soluble in water (and also insoluble ones) can often be examined with advantage in some other medium, such as **Gum** Dammar, Canada Balsam, Glycerine, etc.

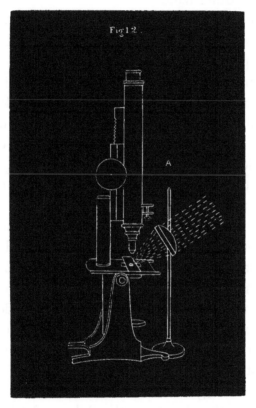

In the Brewery, the substances requiring a high power, such as Yeast, Beer sediments, etc., are generally in a liquid state, and can often be placed on the slide just as they are, or—as it is advisable not to have the field too full of objects—they may be first diluted with water to the required extent. A drop of liquid just sufficient to spread beneath the cover-glass, may after a few trials, be accurately judged, and thus prevent the necessity of wiping off superfluous liquid, always an untidy matter, and requiring considerable care to avoid more or less disturbance to the specimen under examination. When examining this class of object, frequent alteration of the focus is necessary, and the fingers of one hand may be kept on the fine adjustment, whilst the slide, or stage movements can be controlled by the other hand.

Until some experience has been acquired in the manipulation of the microscope, there is a risk, when focussing with high power objectives, of passing the focal point unawares, and driving the objective down with more or less force on the cover-glass and slide : although the spring of the objective socket—where a spring is provided—may prevent an accident, still fracture of a cover-glass and slide sometimes results, accompanied possibly by damage to the objective itself. To obviate this risk, it is better first to gently run the objective down close to the cover-glass, simply taking a view from the side of their relative positions ; then focussing upwards, and away from the cover-glass with the eye applied to the tube.

A useful selection of minute dissecting instruments may be obtained by mounting needles in small wooden handles, and after softening them in a spirit lamp or other flame, grinding cutting edges, and bending to desired shape.

For most Brewery work, a few pipettes or dropping tubes, and some glass rods with somewhat pointed ends, suffice to place the objects on the slides. A neat and light glass rod can be made from quill glass tubing (about $\frac{1}{8}$ in. internal diameter), by the use of a mouth or stand blowpipe, or even with an ordinary gas or Bunsen flame.

A piece of Platinum wire fused into a glass rod is a convenient instrument for work with Bacteria, as it may so readily be sterilised by heating red hot. A small wash-bottle to hold distilled water, or bright clean well-water, is very useful, and may be constructed either from a 3 oz.

narrow-necked bottle, or from a forcing flask (Fig. 13). *a* is the blow tube and *b* the delivery. The same arrange. ment of tubes does for both forms.

The following reagents in small stoppered bottles may also be kept on or near the microscope table :—weak solutions of Ammonia, Iodine and Methyl Violet. The use of these will be apparent later, and their preparation is described in the Appendix under Reagents.

It is highly important that a reliable record should be kept of all objects of interest, for the purpose of reference,

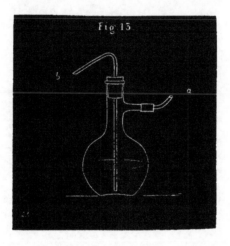

or in cases where any divergence from ordinary appearances in well-known objects is presented. To do this effectually, some little skill and practice in drawing are requisite, and if these are not already possessed by the operator, should be cultivated without delay. Written notes, remarks, and date should be appended to these drawings, which had better be contained in a sketch-book.

The Camera Lucida has already been referred to, as also the Neutral-tint Reflector; for each of these a horizontal position of the microscope is necessary, and this constitutes a marked disadvantage in their employment; illumination, focussing and manipulation of the object being rendered decidedly more difficult ; but on the other hand the outlines and dimensions of objects may be obtained with fair accuracy. In using either of these appliances, the tube of the microscope is laid parallel to the surface on which the instrument stands, so that the vertical height from the centre of the eye-lens to the sheet of paper placed underneath is about 10 inches (Fig. 14). The Camera Lucida is then adjusted till the centre of the field of view lies perpendicularly below the eye, and the image of any

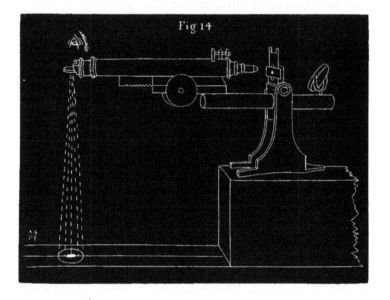

Fig 14

object focussed by the microscope appears superimposed on the paper below.

In the case of the Neutral-tint Reflector which is at a fixed angle, it is generally only necessary to rotate the eye-piece to which it is attached until the image is vertically beneath. Outlines are first drawn with a fine-pointed pencil, and detail filled in, partly with the reflecting arrangement and partly by the unassisted eye or from memory.

A method that we have frequently employed for drawing with the microscope and which answers well, is the following :—a small desk, with its upper surface sloping at the same angle as the stage of the microscope when in use, is placed close alongside the instrument on its right. A piece of drawing paper is pinned on, and the left eye being applied to the microscope, the image is seen by the right eye to overlap the paper; drawing may be carried on continuously, both eyes being applied to the paper from time to time as a check.

In order to agree with any inclination of the stage, the following elaboration of the desk is suggested. The top is made movable by being hinged on to its box on the side nearest the observer, the side farthest from him being provided with curved brass bands, and setting screws

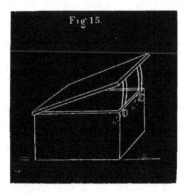

Fig. 15.

for each, so that the lid may traverse an arc of nearly 45 degrees. (Fig. 15.)

It is a good plan to first sketch lightly in pencil, and then to etch with a fine pen and Indian-ink. The drawings may be bounded by a circle, or a stencil-plate may be obtained cut in copper, giving any desired margin.

It may be here remarked that with constant use of the microscope it is a good plan to cultivate the use of either eye.

An appliance is sold by some opticians consisting of an ebonite disc, so fixed in relation to the eye-piece that the unused eye of the observer is shielded and can be kept open.

Micrometer lines which are required for the measurement

of objects and the estimation of magnifying power, are conveniently copied by the Camera Lucida or Neutral-tint
Reflector ; or, the microscope being in its ordinary position,
the left eye is applied to the tube, whilst with the right eye
the observer marks dots corresponding with the magnified
lines of the micrometer, on a slip of paper held in front of
the stage, and then ruling parallel black lines through these
dots makes further comparison.

For the measurement of objects, either the stage—or
eye-piece micrometer is employed, usually the former, the
latter not being of such general application ; it is custom-
arily ruled in hundredths and thousandths of an inch, or
parts of a millimetre. The lines as seen through any of
the combinations of lenses are depicted by drawing as
described, and on or between the lines so obtained the
object is drawn, and its relationship to them determined by
measurement. For instance, were half of the space indicated by $\frac{1}{100}$ of an inch filled, the linear measurement of
an object so filling it would be $\frac{1}{200}$ inch, or were one-third
of a $\frac{1}{1000}$ inch space so occupied, the object would be $\frac{1}{3000}$
inch in size.

The magnifying power of a microscope is not a definite
amount which can be fixed once for all ; it is dependent
upon the condition of the eye of the observer ; but in recent
times it has been customary to calculate the magnifying
power for a distance of ten inches, the average minimum
distance at which objects are distinctly visible to the normal
eye ; so that for the estimation of magnifying power with
the stage micrometer, the slip of paper should be placed
ten inches from the eye ; the lines then drawn being compared with an accurately divided inch scale. Suppose,
for instance, five of the $\frac{1}{1000}$ inch spaces are required to
fill one inch, the power $= \frac{1000}{5}$ or 200 diameters, or if four
of the $\frac{1}{100}$ inch spaces fill one inch the magnification would
be $\frac{100}{4}$ or 25 diameters. Plate I. shows some micrometer

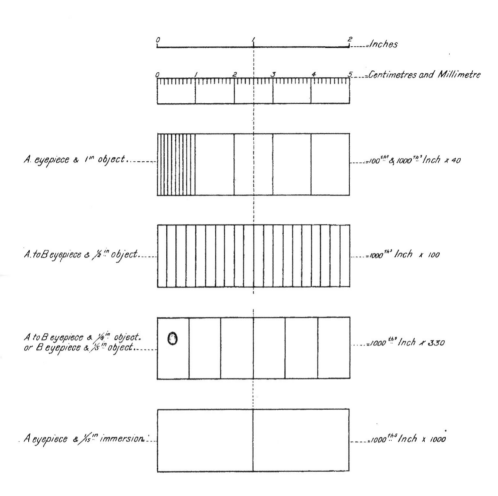

0 ·····················1·····················2 ·····=Inches

0   1   2   3   4   5 ·····=Centimetres and Millimetre

A. eyepiece & 1ⁱⁿ object.·····   ·····=100th & 1000^{ths} Inch × 40

A. to B eyepiece & ½ⁱⁿ object.·····   ·····=1000^{ths} Inch × 100

A to B eyepiece & ⅛ⁱⁿ object.
or B eyepiece & ⅛ⁱⁿ object.·····   ·····=1000^{ths} Inch × 330

A eyepiece & ⅟₁₅ⁱⁿ immersion.·····   ·····=1000^{ths} Inch × 1000

Comparison of Micrometer lines with Metric and Inch Scales

lines magnified with values attached (in one a yeast cell is shown in its relative magnification), also the relation of the micro-millimetre scale to the inch. English measurements are frequently given in $\frac{1}{1000}$ of an inch, whereas foreign measurements are in $\frac{1}{1000}$ of a millimetre or micro-millimetre (μ) as it has usually been called,* and this method is rapidly coming into general use. The relationship of the thousandth of an inch to a micro-millimetre is as 1 : 25.4.

This matter has been entered into in some detail, as it is both interesting and useful to know the magnifying power of the combinations, and the size of the objects under examination.

Of course the most complete and valuable pictorial records of objects are those furnished by the aid of photography, but very few microscopists have, till recently, had time or patience to pursue this branch of investigation, the old methods requiring such cumbrous and expensive apparatus. Some Burton friends of the authors have latterly obtained excellent results with greatly simplified apparatus, and in the Appendix a brief sketch of a suitable and convenient method of Photo-micrography is given.

---

* As Physicists and Electricians have used this word micro-millimetre to indicate the millionth of a millimetre the term MICRON has been suggested to express the thousandth of a millimetre ; and in June, 1888, the word was adopted by the Royal Microscopical Society.

CHAPTER III.

ALCOHOLIC FERMENTATION.

A T various stages in the Brewing process we can, by the aid of the microscope, determine the presence of organisms of various kinds, ranging from those originally present in the materials, to those which are introduced up to the very last by exposure to air. It is in the identification and study of these, that the microscope has, for the Brewer, its more important applications, and especially as regards the organisms added in the form of yeast, which substance, although containing in most cases a preponderating number of a desired species, is seldom free from forms that are undesirable or positively injurious to the process.

Speaking generally, we may include the organisms with which we have to deal in three important classes. First, those provocative of alcoholic fermentation by the breaking up of Sugars—scientifically known as the Saccharomycetes. Secondly, the moulds or Thallophytes, giving rise to objectionable products of decomposition from a variety of substances. Thirdly, the Bacteria or Schizomycetes, inducing changes which are, from the Brewer's point of view, mainly useless or deleterious.

Besides these organisms we occasionally, in the case of Brewing waters and in a few uncommon instances, meet with some of a higher grade in the animal and vegetable kingdoms, but they can hardly be considered as directly affecting the Brewing process.

It is our purpose to consider the forms of life according to the classification indicated above, which represents sufficiently well their position in the scale of life, the lowest organisms being those of Class 1. Our attention is thus first claimed by what for Brewers is the most important group, viz., the Saccharomycetes or alcoholic ferments, which in this chapter we will consider as regards their general character.

In ordinary parlance the term yeast has been applied to the surface scum or sedimentary deposit separated during the fermentation of Wine, Beer, etc., in yellowish or brownish masses, which, amongst foreign substances, contain the organisms corresponding with alcoholic fermentation—the so-called alcoholic ferments—the cells of which compose the greater part of the separated yeast.

For a long time past, yeast has been used to excite fermentation in Saccharine solutions, the yeast accruing from one operation being the starting point of another. Prior to the times of intentional addition of yeast, fermentation took place either naturally or fortuitously in Saccharine liquids, as for instance, in the expressed juice of the grape or of other fruits, and very probably in Saccharine fluids prepared from cereals, as in the case of the Maize beer, or Chica of the Peruvians, and the Cerevisia of the Romans. Even at the present day so-called spontaneous fermentation leads to the production of certain beverages, including some kinds of Belgian Beer.

All Saccharine liquids may be considered as capable of giving rise to the phenomenon of fermentation, and especially the Saccharine liquids provided in nature, such as fruit juices, etc., the character of a fermented product being in any case dependent, of course, on the substances yielding the fermentable extract. As regards Beer, we have to deal with a beverage of relatively small alcohol percentage, a fact which mainly accounts for the greater susceptibility to change that it displays as compared with wine.

If a liquid in an active state of fermentation be filtered, the suspended yeast may be removed ; and if this be completely effected, fermentation ceases. The yeast so removed is capable, as is well known, of starting a fresh fermentation ; or it may be carefully dried off at a temperature below 100° F., and retain its power for a considerable period, but the temperature of boiling water, or a short exposure to a temperature over 130° F., as well as exposure to a very great degree of cold, destroys its fermentative action. Saccharine solutions which have been boiled in flasks in the laboratory, cooled, and exposed to the air, frequently enter into alcoholic fermentation ; but if, whilst hot, the flask be plugged with cotton wool, the liquid contained in it remains practically unaltered. We have thus evidence upon the following points :—

(1) That there is something in yeast which causes fermentation.

(2) That this property of yeast is destroyed by a high temperature.

(3) That the property is associated with particles that may be separated from the fluid containing them by an efficient filter.

(4) That these particles may be contained in the air, and may be separated from it by causing it to pass through cotton wool. Microscopical examination of a drop of yeast shows what the particles in question are. The earliest recorded examination of yeast in this way was made by Leuwenhoek in 1680, when he ascertained that it consisted of small spherical and oval bodies, but failed to determine their nature. About 150 years later, Cagniard-Latour took up the work at the point that Leuwenhoek had left it, and

added thereto the observations that the globules of yeast reproduced themselves by budding, and thereby exhibited properties including them in the vegetable kingdom. After a short period of comparative inactivity in this line of investigation, Schwann, of Jena, and Kützing, of Berlin, independently, and almost simultaneously, re-discovered the facts established by Cagniard-Latour, and so progressively the nature of yeast became pretty clearly known. Since then its life-history and functions have been minutely investigated by Pasteur, Reess, Hansen, and others, and many distinct species of Saccharomycetes have been clearly identified.

Some evidence has already been adduced in this chapter as to fermentation being the result of a *specific* organism. In reality no fact has been more clearly demonstrated by scientific proofs—provided more especially by the beautiful and classical researches of Pasteur—than that without the presence of a ferment cell, fermentation cannot take place ; and that the removal from, or the killing of, the organisms in a fermentable solution suffices to bring any fermentative action to an end. This statement need only be modified to the extent of saying that it has been found by Lechartier and Bellamy, and subsequently established by Pasteur* that under certain conditions an internal fermentation takes place in fruits, accompanied by the production of alcohol. This phenomenon appears to be dependent on some kind of residual vitality in the fruit tissue, and has no connection with the definite ferment cells that were shown by Pasteur to be adherent to the surface of fruit, especially the grape, and capable of fermenting the expressed juice of the same.

Space does not permit us to trace the successive stages of the researches alluded to, nor can we here review the various theories as to the nature and mode of action of

---

* Etudes sur la Bière.   Pasteur, p. 258, et seq.
  Translation of same, Faulkner and Robb, p.266, et seq.

ferments, that stimulated the prosecution of them. Pasteur so completely cut the ground away from under the fanciful theorists of the Liebig School, and showed so clearly the organised nature of the alcoholic ferments proper, that the only immediate point remaining undecided is, as to how the cells exercise their functions ; whether action takes place inside or outside the cell. Many scientists cling to the idea of the diffusible matters of the saccharine fluid passing into the cell, and there undergoing decomposition, whilst others accept the theory enunciated by Nägeli, that the activity of the alcoholic ferments is due to vibrations of the protoplasmic contents, communicated through the cell wall to the substances immediately adjacent to it. The fact that the thinnest membrane intervening between yeast and a fermentable liquid prevents fermentation—as shown by Dumas—is somewhat against Nägeli's theory.

The destructive action, already spoken of, that heat exercises on yeast, applies in a measure to organisms productive of fermentations other than alcoholic. The absence of all organisms and the consequent immunity of a liquid from fermentative or putrefactive change, constitutes a state of Sterility ; the only changes then possible being traceable to other agencies than those under consideration, such as oxidation by air, evaporation, etc., etc. Regarding the yeast cell, then, as the direct cause of fermentation, we will, after the briefest consideration of its chemical composition, proceed to ascertain its appearances during the progress of its fermentative career, contenting ourselves for the moment with the knowledge that we are dealing with an organism consisting of a simple oval or spherical cell, with an outer envelope or cell wall, and internal viscid matter known as Protoplasm.

The organic nature of yeast is easily demonstrated by drying and charring in a silver or platinum dish ; a smell similar to that of burning animal matter is produced, whilst a mass of charcoal and mineral matter is left, which, if completely incinerated, is reduced to a white ash, consisting entirely of mineral matter. Chemical analysis proves that yeast contains Carbon, Hydrogen, Oxygen and Nitrogen, with relatively small quantities of Sulphur, Phosphorus, Potassium, Magnesium, and Calcium. These elements are variously combined to form the constituents of yeast cells, namely, Albuminoid or Proteid matter, Cellulose, Fat, Saline substances and Water. The envelope of the cell contains the cellulose or substance resembling cellulose, and some of the mineral matters ; the protoplasm containing the protein compounds and fat with the larger proportion of the mineral salts.

The above elementary matters must, of necessity, be contained in some form in liquids destined for the production of alcohol with a corresponding increase of yeast. They are contained in malt worts, the sugar Maltose, furnishing Carbon, Hydrogen, and Oxygen, as well as giving rise to the alcohol formed ; whilst the Albuminoids and Amides furnish the bulk of the organic constituents, especially Nitrogen ; and the phosphates and other salts of Potassium, Lime, and Magnesium present in the worts, provide the Inorganic or mineral constituents. As Pasteur has shown, the simplest combination of substances that will sustain yeast is a solution containing Sugar, phosphates of Potash, Lime, and Magnesia, and Tartrate of Ammonia, the latter supplying the required Nitrogen : the power of manufacturing protein from Tartrate or other salts of Ammonia, being a distinct peculiarity of vegetable life.

Yeast usually has a slightly acid reaction, and grows better in an acid than an alkaline liquid. The variation in the conditions of nutriment affecting yeast involve considerations which, being chiefly of a chemical nature, are scarcely in our range. The microscope may well be employed to detect such differences caused in the appearance of yeast, but does not necessarily afford an explanation of them. We shall, therefore, confine ourselves to mentioning at the end of this chapter some general considerations in connection with malt worts, and will now proceed to the microscopical examination of yeast, and the study of the yeast cell as a complete organism. We may prepare some yeast for viewing, as follows :—A clean glass slip is taken, and by means of the wash bottle (Fig. 13) or a pipette, a small drop of clean water is placed in the centre. The sample of yeast, which may conveniently be contained in a small tumbler, is now thoroughly stirred with a pointed glass rod, which, before withdrawal, may be rotated against the sides of the vessel to remove the excess of adhering yeast. The rod is then carefully dipped in the water-drop till a quantity of yeast is introduced sufficient to impart a slight milkiness, when the drop is stirred with the clean end of the glass rod. A cover-glass is now put on, and gently pressed down with the finger-nail, any liquid pressed out being removed with a small piece of blotting paper. Another method of preparing yeast for examination is to stir a small quantity into a wine glass of clean water, and take a drop of this for the slide. We prefer, however, to use the first method.

When viewing yeast for the first time, it is instructive to use an objective magnifying some 40 or 50 diameters, in order to realise what a very minute organism is being dealt with. Afterwards the combination giving about 300 diameters may be used, this being the power generally applicable to the examination of Yeasts, Beer residues, etc., etc. With this degree of magnification we see a collection of cells of spherical, oval or ovoid form. These cells consist of a thin-walled sac or bag, containing a somewhat viscid fluid. We have already spoken of these respectively as cell-wall and protoplasm, their scientific designations.

The cell-wall is an integument which, although thin, has

considerable elasticity and resisting power ; it may however be burst by a sudden shock, such as a blow upon the
cover-glass with the blunt end of a lead pencil ; when the
protoplasm is seen to have emerged from the sac, whose
nature can now be distinctly ascertained.

On examining perfect cells more closely, we become
aware of differences in the nature of the protoplasmic contents ; in some parts of the cell it is decidedly clearer than
in others. These clear portions are called the Vacuoles
(shown in Fig. 16 on diagram scale), and are considered to

Fig. 16.

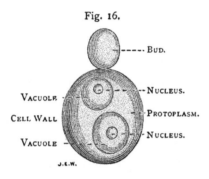

have their origin mainly in the withdrawal of nourishment
from the protoplasm of the parent cell during the reproductive process ; the protoplasm being replaced by transparent cell-juice, as Reess terms it, probably of a more
aqueous nature than the rest of the protoplasm. At the
same time it is by no means certain that the vacuoles
cannot appear and be well marked apart from such action.
We are inclined to believe they can. In any case their
appearance is very much influenced by varying conditions
of temperature and aeration where reproduction does not
take place.

Healthy yeast cells usually show at least one vacuole,
but often two, and sometimes three. Inside the Vacuole
may be seen small granules, and these not unfrequently
move actively in the clear protoplasm. We have known a

"Stone Square" yeast and specimens of London yeast show this very plainly. The granules are called Nuclei (Fig. 16).

On examining full-sized, well-vacuoled cells carefully, a dark spot may be frequently detected, which on causing the cell to shift its position by touching the cover-glass, pressing it slightly, or in some other way imparting movement, is seen to lie on the cell wall. This is the point at which the bud is appearing. This bud or young cell (Fig. 16) gradually enlarges, drawing its supply of nourishment from the parent cell till it attains about the same size ; the area of attachment then decreases till the young cell becomes detached from the parent, and goes forth fitted to procure a living for itself. There is no doubt that the same cell can bud more than once, and in some cases at more than one point of its envelope at the same time. Pasteur, by continued observation of a budding cell, found that at the end of two hours it had produced six daughters, or that the two (*i.e.*, the cell and its bud) had become eight cells. The enormous rate of multiplication thus indicated does not, of course, take place in the Brewery, where a comparatively large excess of yeast is employed. Still the reproduction is considerable.

The functions of the Vacuoles and Nuclei is not understood, but they may be taken as indicative to a certain extent of the age and degree of activity of the cell. Take, for instance, some of the yeast thrown up last of all in a skimming fermenting vessel, or the last part of the spurge from unions or cleansing casks. It consists very largely of new cells which show no differentiation of the protoplasm, that is to say, no vacuoles or nuclei. They are homogeneous, and possess an appearance of uniform semi-transparency. Plate II., Fig. 1a. Such cells are the recently-formed ones, and cannot be considered as having arrived at the full degree of activity, which we believe we are correct in

Fig 1

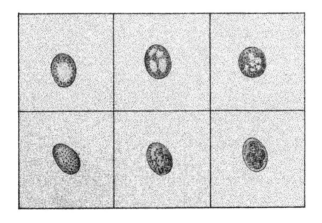

Development of Yeast

Fig 2

Sporulation of Yeast
(after Reess)

ascribing to the completely differentiated ones. After a rest of a day or two these new cells begin to show vacuoles (Plate II., Fig. 1, b and c), and in a fresh fermentation become parent cells, d, e, f. These phases of development are progressively represented in the figure, the cell arriving at its full maturity, and reproducing its typical form by budding. During every normal fermentation a small proportion of the cells deteriorate and die, owing probably to natural decay, after passing through several fermentations with repeated reproduction. The growth of yeast under unfavourable conditions of nutriment, temperature, &c., will increase the proportion of dead cells, besides weakening both old and new cells.

The progressive changes in appearance following the separation of yeast from the fermented liquid are, under ordinary conditions of storage, an apparent thickening of the cell-envelope, the enlargement of the vacuoles, or the coalescing of all vacuoles into one, accompanied by a greater sharpness about the nuclei (see Plate II., Fig. 1 g).

The next stage is that the whole protoplasm begins to grow dense, and takes on a speckled appearance, the cell lessening in size (h).

The cell shrinks still further from yielding up of cell-sap, the protoplasm becoming quite granulated (i), and frequently showing a slight yellowish green tint. The final diminution of the cell is to about $\frac{1}{2}$ or $\frac{2}{3}$ its original diameter, and at this stage it is probably dead, for it is difficult to give a precise point in the decadence at which the cell is incapable of revival; it being certain that yeast cells apparently dead are sometimes only torpid, and may begin to grow and give rise to fresh fermentation in a stimulating liquid. For all practical purposes we may, however, regard the small granulated cells in any sample as useless. A very good method of distinguishing between living, and dead or torpid cells, is based on the resisting power that living cells exhibit to dyes. It consists in running a little solution of methyl violet or carmine at the side of the cover-glass, and causing the dye to traverse the yeast under examination by applying a small piece of blotting paper to the opposite side. The proportion of cells stained is noted. If the dye solution be weak the active cells do not stain, the torpid cells stain slightly, whilst the dead cells show a deep colour. Weak iodine solution may be used for the same purpose, the dead cells staining brown. A somewhat more convenient mode of manipulation is to put a drop of dilute stain or reagent on the slide and stir the yeast into it.

Yeast is capable of reproducing itself in another very interesting manner by the formation of internal spores, known as Endospores or Ascospores, the process being termed multiplication by Endogenous division. The ordinary forms of Saccharomyces seldom or never exhibit this phenomenon in fermentation. It is induced by withdrawing the yeast from fermented liquids, and exposing in a moist condition on a plate of some porous material such as Plaster of Paris, or on slices of potato. Frequent examination is made with the microscope, and where the phenomenon occurs, the protoplasm is seen to gradually separate into a number of parts, usually four, each of the portions approaching a spherical form, and becoming surrounded by an envelope of its own. (Plate II., Fig. 2.) As these ascospores reach their maximum size, the old cell wall gives way and they escape, and in a nutritive solution are capable of budding, giving rise to ordinary yeast cells again.

The first communication on this subject was made by De Seynes in 1868. He observed ascospore formation in Mycoderma Vini. Reess's work followed, describing ascospore formation in several kinds of Saccharomyces. The better method of cultivation, viz., on plaster blocks, was introduced by Engel, and other investigators have contributed points of information. Hansen, however, has made the subject his own, and put the ascospore formation on a well defined basis as regards conditions of time, temperature, etc. He ascribes the failure of other investigators in obtaining ascospores, to the fact that only young and vigorous yeast sporulates. He has obtained ascospores from English yeast as well as Foreign yeast of various kinds, and makes use of the ascospore formation to determine the identity of any given yeast. In another chapter we shall go somewhat further into this matter.

Having thus acquainted ourselves with the chief facts relating to the life history of yeast, we will conclude this chapter with some general considerations regarding influences obtaining in the Brewery, which tend to modify the microscopical appearance of the ferment. Speaking generally, Low gravity malt worts may produce clean and uniform looking yeasts, but they are seldom vigorous ones.

Malt worts of medium gravity—other conditions being favourable—produce the most uniform yeast.

High gravity malt worts produce a vigorous yeast, but not of uniform type, and capable of greater fermentative than reproductive power.

The foregoing remarks apply also to cases where a moderate proportion of Brewing sugar is used.

Another set of conditions influencing the activity and the microscopic appearance of yeast, is the range of temperature through which it has been taken in the fermentation. Taking into consideration all known kinds of yeast, we find that the phenomenon of fermentation is exhibited at all temperatures between 32° F. and 130° F. but for the majority of species, temperatures ranging from 54°—75° F. are the most favourable. The pitching temperatures of the United Kingdom (distillers' fermentations excepted) approximate to the lower figure, and the maximum temperatures attained, are some degrees below the higher. Now

confining ourselves to the range of temperature employed
in English breweries, apart from other modifying influences
of greater and less importance, we believe that low temperatures, as a rule, produce the cleanest yeast but not the
most active.   High temperatures tend to produce an unclean yeast of considerable activity, but liable to rapid
deterioration.

Intermediate temperatures produce a good type of yeast
and of reasonable cleanliness.   Every system of Brewing
has its own most favourable range of temperature for fermentation, discoverable from the character of the yeast and
the nature of the Beer produced, other conditions being of
course adapted to a favourable result.

Another factor of considerable importance, influencing
the activity and consequently the appearance of yeast, is the
degree of Oxygenation of the worts ; that is the amount of
oxygen dissolved or held in loose combination at the time
the · worts reach the fermenting vessel.   Now, although
yeast can live and increase in saccharine solutions without
free oxygen, yet it is only to a limited degree, and for
continued and vigorous fermentation, oxygen is absolutely
necessary.   During the apparent quiescence of the yeast
for many hours after pitching at a normal temperature, the
dissolved oxygen is being absorbed by the living cells, and
on the supply being practically exhausted, the yeast in-
vigorated to a corresponding extent, commences to attack
the saccharine matter, and the production of alcohol and
carbonic acid gas commences, the yeast meanwhile entering
on the stages of its reproductive career, which with the
assimilation of nutrient matters from the wort, goes on till
the fermentation ceases.   A stock of fresh cells is provided
far in excess of those which naturally expire in the process,
the new cells retaining under normal conditions the characteristics of their progenitors.

Returning for a moment to aeration or oxygenation,

nothing tends more to the production of a feeble deteriorated yeast than insufficient oxygenation of worts, and under ordinary Brewery conditions (*i.e.*, in absence of special aerating apparatus), worts cannot very well be over-aerated.

In a normally attenuated beer brewed with a clean and good type of yeast, a slow fermentation goes on in cask, which, when caused by ordinary yeast or Saccharomyces Cerevisiæ, seldom gives trouble, but unfortunately for the Brewer the secondary fermentation is not unfrequently set up by strange ferments, better able to exist under the obtaining conditions than Sacc. Cerevisiæ, especially where the original worts were not of proper character. This is the cause of so-called Fret, Sickness, etc., which are described in connection with the particular ferments associated with the changes.

Summarizing the action of yeast in malt worts as regards its life history, we then have—

First ; the time of rest or of no visible signs of fermentation. During this period the yeast absorbs oxygen, and commences its vegetative and assimilative functions.

Secondly ; the period of increasing activity, the yeast rapidly attacking sugar, the formation of new cells progressing simultaneously.

Thirdly ; the slackening and ending of the primàry fermentation during which the active yeast is mainly conveyed to the surface.

We advise the student to carefully examine specimens of yeast taken at various stages of fermentation and cleansing.

CHAPTER IV.

ALCOHOLIC FERMENTS OF THE ENGLISH PROCESS.

THE researches of Pasteur were unquestionably the starting point of all important investigations in connection with fermentation, and they may be considered as having largely contributed to elevate Brewing to a scientific industry; for although the microscope had been employed in Burton and perhaps elsewhere in the Kingdom, for the selection of yeast, Brewers, as a body, awoke to the fact that the condition of their yeast was of chief importance as determining the production of a satisfactory beer. The main point of Pasteur's researches was the indication of the danger to be expected from organisms capable of producing Acetic, Lactic, and Butyric acids, and other objectionable products; and the means of removing such risks or reducing them to a minimum. At the same time he by no means lost sight of the possible influence of strange forms of Saccharomyces; but doubtless at the time, this did not appear to him the chief issue involved. His work, we need hardly say, marks a distinct epoch in Brewing science.

It will be plain, from what has been already said, that the term *yeast* is indefinite, as one or many things are covered by the same term, and the word *ferment* alone is not any more precise, including as it does at present, organized ferments such as the Saccharomycetes, Moulds, and Bacteria, besides substances having a curiously specific action, such as Invertin and Diastase, which are unorganized ferments. Despite its occasional impurities, we may regard Brewers' yeast as consisting of the cells of a living vegetable organism capable of decomposing the saccharine matters existing in worts, forming therefrom Alcohol and Carbonic Acid gas with a small percentage of what may be considered as bye-products, *e.g.*, Glycerin, Succinic Acid, etc.

After these preliminary remarks, and having already described the life-history of the yeast-cell generally, we may now address ourselves to a consideration of the definite alcoholic ferment forms exhibited by the beers brewed in the United Kingdom. So far as our present knowledge goes, we have to deal with the following species :—

Saccharomyces Cerevisiæ,

Pastorianus,

Coagulatus,*

Ellipsoideus,

,, Minor,

and one or two species of rarer occurrence,

Saccharomyces Apiculatus,

,, Exiguus,

besides certain organisms which may at times act as alcoholic ferments, viz. :—

Mycoderma Vini,

Mucor Racemosus (ferment form).

The different yeasts in use throughout the United Kingdom, vary so considerably as regards appearance and degree of activity, and give rise to Beers of such essentially different character, that mere modification of one species of

---

* We suggest this as a convenient title for the Caseous ferments, of which there are probably two or more varieties.

Saccharomyces by differing conditions of the process, would seem to afford a quite inadequate explanation of such diver-gencies, and it is more than probable that different varieties, if not species, of Saccharomyces Cerevisiæ are in use at the present time.

Failing more exact knowledge, however, we may for the present regard the large proportion of cells in a clean and active yeast as being those of Saccharomyces Cerevisiæ, and we may well acquaint ourselves first with the ap-pearance of yeast grown under different conditions, but relatively pure in respect of the absence of Bacteria, and obviously unusual or "wild" forms of yeast. Plate III., Fig. 1 shows a typical sample of Burton yeast of a high degree of purity and fermentative vigour ; the features worthy of observation are :—The uniformity of the shape, size, and appearance generally. The tendency to an elliptic or ovoid form. The clearness of the vacuoles ; and lastly, the absence of extraneous matters.

Plate III., Fig. 2 represents a specimen of London yeast of good. quality. It will be noted that there is not the same regularity of size and shape ; that the cells are throughout larger ; and the internal features vary to a not inconsiderable extent. In some cells the vacuoles are unusually large, and the nuclei very distinct. We may here remark that, besides differences in the character of nutriment, the higher ranges of temperature of fermentation tend to increase the size and diminish the uniformity of yeast cells. There is probably a relationship between these facts and the appearance of London yeast, of which the plate represents, we consider, a typical specimen.

The other characteristic yeasts of this country are, we should say, the Scotch slow yeast and Stone Square yeast, both of which, considered from the point of view of their appearance under the microscope, occupy a position inter-mediate between London and Burton yeast. They are

PLATE III

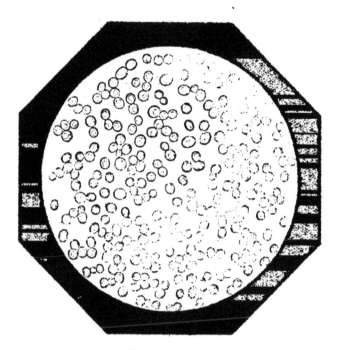

Fig 1 Burton Yeast

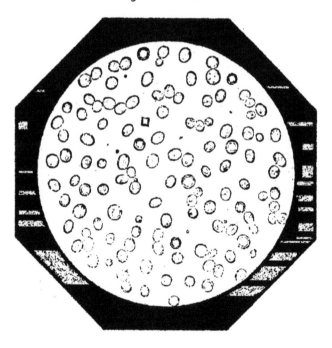

usually somewhat irregular, and generally well-vacuoled, especially the "Stone Square," in which the nuclei are often remarkably distinct and large.

The average diameter of the cells of ordinary Brewery yeast is about $\frac{1}{3000}$ inch or 8 micromillimetres (μ).

The microscopical characteristics which, as a rule, denote an active and healthy yeast, are the following :—Uniformity of size and shape. Sharpness of cell outline, indicating a strong cell-wall. Presence of vacuoles, which should be clear and of fair size, neither large nor small; for if the vacuoles are barely perceptible the yeast is probably too young for use, and if the vacuoles are large it is a sign of exhaustion by much previous reproduction, or unfavourable conditions attending its growth. The nuclei should not be too plain, though as already mentioned, some yeasts show nuclei much more distinctly than others, so that due allowance must be made for the particular process by which the yeast has been produced. There should be no noticeable proportion of dead or granulated cells, or foreign matters such as strange yeasts, filamentous bodies, "grounds," or dirt of any kind. To emphasize these remarks, and in contrast to the plates already shown, two fields of yeast are represented in Plate IV., Fig. 1 being the Burton yeast of Plate III., Fig. 1, in a deteriorated and granulated condition, quite useless for pitching; whilst Fig. 2 shows an exhausted yeast, accompanied by an excessive amount of extraneous matters.

As we proceed, we shall enter more fully into the various causes of yeast deterioration, and the appearances in connection with the same ; at present it is sufficient to speak of the signs of degeneration traceable in the cells alone · they are the following :—

Unusual thinness or thickness of cell-wall, more especially the former.

Abnormal clearness and largeness of the vacuoles.

Very distinct nuclei (except under certain conditions mentioned).

Speckled appearance of the protoplasm.

Smallness of fully matured cells (except in the case of yeasts from very strong ales).

Large thin-walled cells.

Irregularity in size and shape.

And lastly, any marked percentage of dead or torpid cells.

We strongly advise the student to examine yeasts from all sources open to him, and to draw typical fields in a suitable book, appending notes to each specimen as regards its source and age, and any marked peculiarities of appearance. It is true it requires much practice to draw yeast successfully, but it is worth the trouble, for if the drawings are not masterpieces, the mere attempt to make them will impress certain facts on the mind, when otherwise there would probably be no lasting recollection of the specimen. A hard pencil and smooth drawing-paper are sufficient, but if desired, the pencil sketches may be rendered into pen-and-ink etchings, though this is even finer work than pencil drawing.

We may now consider the alcoholic ferments other than Sacc. Cerevisiæ, commencing with

SACCHAROMYCES PASTORIANUS.

This name has been conferred on a ferment which was first identified by Pasteur in the secondary fermentation of wine, and later, of Beer.

Reess also noticed its presence towards the close of vinous fermentation. He adduces some not very strong evidence as to its inability to ferment cane Sugar.* Grape sugar would appear to be easily fermented by it.

---

* Untersuchungen über die Alcoholgahrungspilze.—Dr. Max Reess, p. 29, foot note.

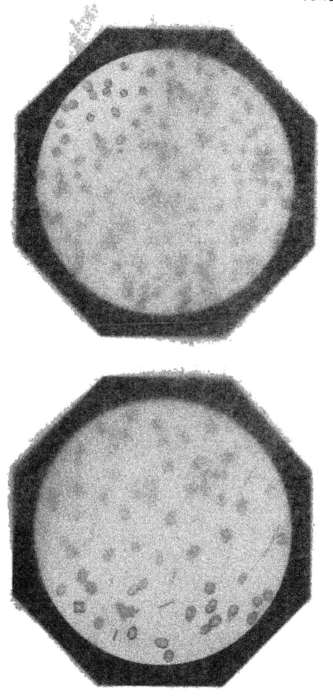

PLATE IV

Unsaturated Yeast

× 300 About

Wm. Newman & Co. Sc

PLATE V.

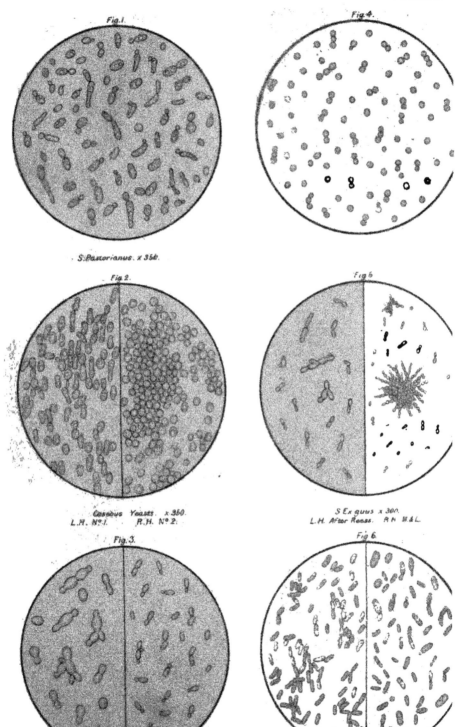

Fig. 1.

S. Pasterianus. x 35b.

Fig. 2.

Casebus Yeasts. x 350.
L.H. N°.1.   R.H. N°.2.

Fig. 3.

S. Ellipsoideus. x 350.
L.H. After Reass.   R.H. N°.4.

Fig. 4.

Fig. 5.

S. Ex guus. x 300.
L.H. After Reass.   R.H. N°.&.L.

Fig. 6.

Mycoderma Vini. x 300.
L.H. Kerahton form.   R.H. Submerged form.

West, Newman & Co. Sc.

Plate V., Fig. 1 represents the ferment as it is generally seen in English Beers. The peculiar elongated cells will be noticed, which bud towards one end of the longer axis and generally on one side of it. The cells of S. Pastorianus, when growing freely and in a state of purity, show a tendency to become a shorter ellipse; the vacuoles being very distinct.

With a ferment so variable in size, it is difficult to give representative dimensions; the shorter axis is about 4—6 μ long, whilst the longer axis may attain 18—22 μ under ordinary conditions.

S. Pastorianus is a common cause of the secondary fermentation of English Beers, and doubtless frequently plays a part, as an accompaniment of several other ferments. It has been shown by Brown & Morris* that S. Pastorianus and also S. ellipticus—unlike ordinary S. cerevisiæ—are able to ferment Malto-dextrin. When growing alone, or in large proportion, S. Pastorianus may give rise to very troublesome " frets ; " for often comparatively little gas is formed in relation to the number of cells visible, and these last have, in the earlier stages of their development, a great tendency to remain suspended in the Beer, owing probably to the slight specific gravity of the cells. Its growth seems to be facilitated by the presence of an undue amount of dissolved oxygen in finished ales. We have frequently seen samples of ale, taken in bottles about the time of racking, develop an active S. Pastorianus fermentation within 36 to 48 hours, the quantity of cells being remarkable, but the gas-evolution comparatively slight. The ferment may be often found in the deposit of bottled beers that are fit for consumption.

" Forced " Beers not unfrequently show S. Pastorianus, and, as in the case of other kinds of Saccharomycetes similarly developed, a much greater variety of form is

---

* Journal Chem. Soc. Trans. Vol 47, 527.

exhibited, owing probably to the elevated temperature (80° F.) of the Forcing Tray.

As regards the conditions of the Brewing process that give rise to S. Pastorianus, we can only say that the following are probably amongst them :—

Impure state of the " Store" yeast as regards foreign ferments or " wild yeast."

Insufficient attenuation and yeast production.

A combination of fineness and flatness at Racking in conjunction with inadequate attenuation.

Although a " Pastorianus fret" is frequently accompanied by a distinctly unpleasant smell and flavour—more especially the former—still the fret may pass off and be succeeded by a normal fining with gas production and disappearance of the unpleasant accompaniments of the S. Pastorianus growth. Frets induced by other forms of Saccharomycetes occasionally pass off in a similar way.

Dry hopping undoubtedly introduces many "wild" forms of yeast, and probably amongst them S. Pastorianus, so that with a predisposition on the part of the ale to nourish this particular ferment, its growth readily follows, especially when the ale being racked very bright, S. Cerevisiæ is in deficiency.

THE CASEOUS FERMENTS.

In the course of his researches* Pasteur encountered a kind of Saccharomyces whose cells showed a curious tendeney to agglomerate and form a curdy or cheesy mass. He obtained it from Burton yeast by exposing a nutrient liquid set with this last, to a temperature of 122° F. for 1 hour. A ferment survived this treatment, and exhibited cells closely allied in form to the elongated ones of S. Pastorianus, giving as well, shorter club- or pear-shaped

---

* " Studies on Fermentation," trans. by Faulkner and Robb, p. 200 *et seq.*

cells. Plate V., Fig. 2, left half disc, is a drawing of what we believe to be an identical yeast which we have seen in Bottled beer deposits and Forced ales. On cultivating his caseous yeast in an artificial medium—Raulin's fluid (for composition see Appendix C)—Pasteur found that oval and spherical cells were developed, but on restoring the ferment to worts, the original irregular and pear-shaped form re-appeared.

The year before last, one of us described in a paper before the Laboratory Club,[*] a caseous ferment obtained from Burton yeast, which seems to be dissimilar to that described by Pasteur ; for when grown time after time in Bitter wort, it consists of spherical cells [Plate V., Fig. 2, right half disc], of $\frac{2}{3}$ to $\frac{4}{5}$ the size of ordinary S. Cerevisiæ, and under no condition assumed the pear-shaped form, although quite 30 or 40 lbs. weight of the ferment must have been handled in the various experiments made with it. For the present it may be convenient to call this ferment Caseous Yeast No. 2 ; and Pasteur's, No. 1.

The experiments with Caseous Yeast No. 2 led to the following conclusions :—

That the ferment possesses a degree of activity hardly inferior to S. Cerevisiæ at from 60° to 70° F. Above 70° F. its activity appears to be greater and it suffers less deterioration than S. Cerevisiæ. Attenuation also goes further, owing probably to the breaking up of Malto-dextrin. It acts mainly as a bottom ferment, very little going into suspension, except at elevated temperatures, when it has a tendency to break up. A curious feature about the yeast is that it takes up little or no resin from hopped worts, and thus, leaving all the more in solution, produces a beer with a marked resin bitter. · An increased production of acid is also a feature, the normal acidity

---

[*] Transactions of the Laboratory Club, Vol. 1, pp. 32 and 33. "Some of the causes of the deterioration of Brewers' Yeast," by C. G. Matthews.

being nearly double that of ordinary sound ales. The Beer has something of a "Lager" flavour, is very stubborn in brightening, and the keeping properties are of an inferior order.

It will be obvious from these facts that the presence of this ferment in quantity in pitching yeast would be a distinct disadvantage, as a peculiar harsh flavour, a yeasty or resinous bitter and a cloudy beer, would be not unlikely consequences. Mixtures of one-half Caseous yeast No. 2, and one-half Burton pitching yeast, when fermented with a rich wort gave decidedly unsatisfactory beers ; and probably a much less proportion of the Caseous Yeast would do the same.

We believe these Caseous Yeasts to be of much commoner occurrence than is generally supposed ; for instance, a specimen of Lager Beer yeast sent to us by a friend, and said to be derived from an originally pure selected yeast, assumed a strongly-marked caseous habit after putting it through one fermentation at 60° to 65° F., and retained the same so long as it was cultivated. Caseous yeast No. 2, is one of the Saccharomyces forms found not unfrequently on dry hops ; it may often be seen in Bottled ales, Forced samples, and even the deposits of Racking samples sometimes give indications that lead one to suspect its presence : we allude to the agglomerating tendency of the cells.

The diameter of the cells of No. 2 is some 5—6 μ. An average for the long axis of the forms of Pasteur's Caseous ferment is about 10 μ. We have good reason for believing that there are several varieties of this ferment, for which we suggest the name Saccharomyces Coagulatus, as being much more in accordance with the properties of the ferments, than the term Caseous yeast.

SACCHAROMYCES ELLIPSOIDEUS.

This ferment—sometimes called S. Ellipticus—was first noticed by Pasteur, and subsequently described by Reess in connection with the alcoholic fermentation of wine, in which it is of common occurrence. It is by no means uncommon in Beers, and is easily distinguished from S. Cerevisiæ, but perhaps with more difficulty from S. Pastorianus. Plate V., Fig. 3 shows the uniformly elliptic cells, one half the disc—the left hand—being a reduction of Reess's drawing of the ferment, whilst the other half represents it as we are accustomed to see it in ales. Long axis about 6 μ, short axis 2.5 μ. The much greater size of Reess's cells may either indicate, a distinct variety of the ferment; the effect of a more congenial nutrient solution; or a different appreciation of the magnifying power of a microscope.

The smaller form of S. Ellipsoideus, the one which more particularly concerns us, is not unfrequently found associated with cloudy "frets" and "sickness," and in such cases may grow pretty freely. It is probable that Beers brewed from very hard waters are more open to this form of secondary fermentation than others. With the first-mentioned ales a decided "stench" often comes on, owing to the production of Sulphuretted Hydrogen, and from this traces of sulphur alcohols, a strongly unpleasant smelling class of substances. These effects wear off however, and a beer which has been through a very bad Ellipsoideus fret may, if otherwise sound, become quite palatable. S. Ellipsoideus is frequently met with in bottled ales and "forced" samples, but is not always accompanied by smell, but so far as our experience goes a bad stench is generally accompanied by Ellipsoideus. There is some evidence[*] that S. Ellipsoideus tends to impart a vinous scent and

---

[*] Claudon & Morin. Compt. Rend, 104, pp. 1109—1111.

flavour to Beer wort, and an acidity of about twice the normal amount. It has been lately suggested that S. Ellipsoideus should be utilized for the production of a vinous unhopped beer or Barley wine; but we question whether the taste of the public is in accordance with a beverage of this description.

SACCHAROMYCES MINOR.

Under this name, Engel describes a ferment obtained by him from leaven of flour, and to which the leaven owes its activity. It consists of budding cells of a globular form, the diameter of the largest being 6 μ. In Pasteur's fluid it produced only a slow fermentation. The cells placed under favourable conditions sporulated.

A ferment form corresponding with Engel's is sometimes met with in Beers [Plate V., Fig 4], and we may conveniently consider this to be S. Minor. Average diameter 3—4 μ. A free growth in beers under ordinary conditions is rare. We have seen it in Racking beers to the extent of 1 to 2 per cent. of the cells present, and in larger quantities in forced ales. Also in a few cases of fret accompanied by persistent cloudiness with flatness. When present in Racking samples it is usually to be detected amongst the Store yeasts. Like other ferments, it is quite possible it may occasionally be introduced with dry hops. Beyond the fact that its presence in ales denotes yeast impurity, and probably high finishing temperatures in fermentation, either in the "Square" or by subsequent rise of temperature in cleansing vessels, there is little to be said.

SACCHAROMYCES APICULATUS,

Described by Pasteur and Reess as adherent to fruits, *e.g.*, grapes, and associated with vinous fermentation, has

been made the subject of a special research by Hansen.*
It presents the appearance depicted in Fig. 17, being
pointed at each end. It measures from 4·5—7 μ largest
diameter, and 2—3 μ wide. It has been seen in Belgian
Breweries where spontaneous fermentation is employed, and
to a limited extent in other foreign beers ; but according
to our experience is hardly ever met with in English beers,

Fig. 17.

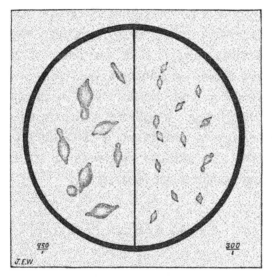

SACCHAROMYCES APICULATUS.

though occasionally doubtful specimens of it occur in forced
samples.

Hansen's observations of the ferment contain much that
is interesting. He found that during the winter, the cells
of the ferment were resident in the earth, underneath the
shrubs on which it appeared during late summer and
autumn, and that the appearance of it was more especially
on certain fruits as they ripened—plums, cherries, goose-
berries, etc.† The descent of rain, or falling of the ripe

* Meddelelser fra Carlsberg Laboratoriet. Tredie Hefte, 1881.
† Pasteur had previously found that ferment cells appeared on the grape at the time of
ripening.—Etudes sur la Bière.

fruit to the ground, caused the cells of S. Apiculatus to become intermingled with the soil, in which it rests during the intervals between its appearance, retaining its vitality until such time as it is liberated by the drying up of the soil and its dispersion as dust.

Hansen shows that S. Apiculatus does not always bud into the typical forms, but may give rise to distinctly oval cells, and also irregular or abnormal forms according to the conditions of nutriment, which when favourable, determine the production of a large proportion of apiculated cells. It ferments Beer-wort feebly, acting as a bottom ferment, and never producing more than 1 per cent. of alcohol, when S. Cerevisiæ would produce 6 per cent. It does not contain Invertase, and consequently is incapable of fermenting Cane Sugar. Amthor,* who has also investigated this ferment, agrees with Hansen as to the alcohol production in worts, and adduces evidence as to its fermenting dextrose more easily than Maltose.

SACCHAROMYCES EXIGUUS.

Observed by Reess in fruit juices in a state of fermentation. Cells about 5 μ long, by 2·5 μ in width at the larger end; it multiplies by budding and sporulation. Reess's form of this ferment, slightly reduced, is shown in the left half disc, Plate V., Fig. 5.

We have on a few occasions seen a ferment in English beers, accompanying a cloudy fret, that we believe to be S. Exiguus; also in one or two samples of bottled pale ale. Plate III., Fig. 4, right hand half disc, represents the forms observed. The marked difference in size between our form and Reess's may be due to conditions attending the growth. It probably excites only a feeble fermentation in

* Zeits. f. Physiol. Chemie, 12, 558.

Beer-wort. The cloudiness in the ales spoken of lasted a very long time, accompanied by marked flatness.

MYCODERMA VINI.

Most Brewers are familiar with the white greasy film that this organism develops on Beer that is spilt about, or left exposed to the air in shallow vessels, as also with its appearance round the taps and shives of casks. In this capacity it is acting as an aerobian ferment, absorbing oxygen, and simultaneously destroying the alcohol of the beer, forming from it Carbonic acid gas and water. Its growth under these conditions is very rapid : Engel implies that in 48 hours, one cell may produce 35,000. Plate V., Fig. 6, left half disc, shows the aerobian form. Its dimensions are very variable, depending greatly on the conditions attending growth. If these are unfavourable the cells may be as small as S. Ellipsoideus, whilst on the other hand in a very free growth they may be quite twice as large. The average length of those depicted is about 9 μ.

Pasteur showed, in addition to other points connected with Saccharomyces Mycoderma or Mycoderma Vini, that when submerged in a fermentable liquid it acts as a slow ferment, forming alcohol and Carbonic acid gas, the cells meanwhile undergoing some alteration in appearance, [Plate V., Fig. 6, right half disc] the filled out brighter cells being the active form. Where ales are badly bottled, and left standing upright with leaky corks, a film of Mycoderma Vini not unfrequently forms on the surface, and may do this to a limited extent where the corks are fairly sound. This film on being submerged can, if the leakage stops, act as an alcoholic ferment. If the leakage goes on, the Mycoderma Vini falls in flakes through the liquid, which becomes utterly spoilt.

Ordinary well-bottled ales **not** unfrequently show some

cells of this ferment amongst the deposit, and there is no particular reason for believing that a small quantity does any harm. The residues of forced beers show it in the same way.

Casks returned to the Brewer only partially filled with ale, frequently include a copious growth of Mycoderma Vini, resulting from exposure to the air. Owing to the peculiar clinging nature of the film, it becomes a question whether ordinary cleansing perfectly effects its removal.

Reverting to the fact that the usual habit of Mycoderma Vini is that of an aerobian ferment, that is to say, it is favoured most by free contact with air, and that growth apart from air is the abnormal state ; it does not seem out of place here, to remark, that Pasteur has shown* that some of the true alcoholic ferments may pass from the anaerobic state of existence to the aerobic. This is often seen to occur in experimental fermentations in flasks, etc. After the primary fermentation is over—the liquid being preserved in a state of quiescence—a film forms on the surface of the liquid and the sides of the flask, consisting of cells of Saccharomyces formed in free contact with air. On submerging these cells in a fresh nutrient solution, the phenomena of active fermentation are reproduced.

Mucor Racemosus.

Having on more than one occasion obtained evidence of the presence of the spores of this mould in samples of pitching yeast, we may here briefly allude to the fact that the mould may give rise to a well-defined ferment form [Plate VIII., Fig. 4], producing cells of very variable size. We have never seen one of the larger cells of Mucor Racemosus in either yeast or beer, but as it is a somewhat feeble alcoholic ferment, it would be at such a disadvantage

* " Studies on Fermentation," trans. Faulkner and Robb, p. 208.

PLATE VI.

Fig. 1.

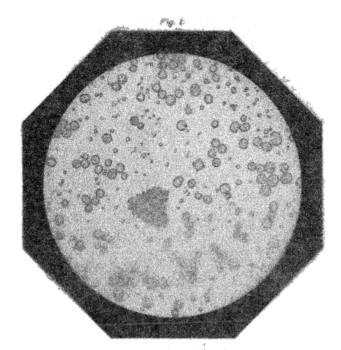

Beer Deposit (Pure)

Fig. 2.

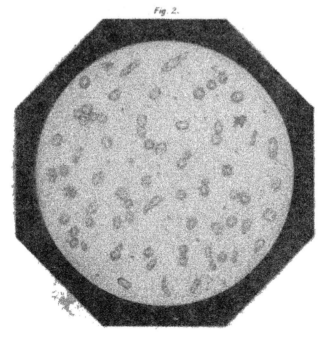

Beer Deposit (Wild Yeasts &c.)

in an ordinary fermentation, that its more luxuriant forms would probably not be produced.

RACKING BEER SEDIMENTS.

Having, as we hope, dealt in a sufficient manner with the forms of Saccharomyces that are, or may be, encountered in the Brewing process as carried out in Great Britain ; some considerations on the deposits thrown down by samples of beer taken at racking will bring this chapter to a fitting conclusion. In the first place, as to the mode of obtaining average samples of the "Brewing." If not taken from a racking square or flattening vessel, but from union or cleansing casks, they should be gathered from more than one vessel, by boring, or from a little sample tap half way up the cask heading ; care being taken as to the exclusion of extraneous matters, *e.g.* borings, in the sample. A thoroughly clean 8—10 oz. stoppered glass bottle is a proper receptacle, and if not perfectly dry, it may be rinsed out with a little of the beer that is being sampled. After standing some 10 or 12 hours, a more or less complete deposition of the suspended matters will have occurred. Some of this sediment may be removed by a glass pipette, or the ale may be run off, leaving a minute quantity of liquid, which by shaking, may be caused to incorporate the whole sediment. With either mode of treatment, a drop is placed on a slide, a cover glass put on, and a microscopic examination made. Plate VI., Fig. 1, shows a normal clean deposit, the extraneous matters being minute spherules, or agglomerations of hop resin, with a crystal or two of oxalate of lime. It may be mentioned that spherules or globules of resinous matter and hop oil, not unfrequently exhibit a peculiar class of movement quite distinct from the vital movements of certain organisms. The student may easily produce a good example of this phenomenon by mixing a little gamboge with water and examining it under the microscope ; minute particles and globules exhibiting much activity will be seen.   This is known as the Brownian movement.   Another point worthy of consideration is, that there is occasionally a possibility of mistaking globules of hop resin for small forms of Saccharomyces and other organisms.   Where there is a doubt, it may be dispersed by treating some of the Beer sediment with a little weak Ammonia or other slightly alkaline liquid, when the resin dissolves ; generally clearing the field to such an extent that a further examination indicates very precisely what organisms or forms are really present.   Plate VI., Fig. 2, represents a sediment resulting from a beer brewed with very impure yeast, in respect to wild forms : some of the ordinary extraneous matters are also given.   In the case of isolated cells, it is by no means easy to refer them to their precise species of Saccharomycetes, though sometimes there is little doubt as to what they are.   In the present case [Plate VI., Fig. 2] S. Pastorianus and Caseous Yeast No. 2, may be recognized, but the usual uncertainty attends the other forms.

CHAPTER V.

Recent Researches in connection with Lager-Beer Yeast, etc.

THE two chief varieties of alcoholic ferments included under the title Saccharomyces Cerevisiæ are, as most of our readers will be aware : first, the ferment which may be considered common to the Breweries of the United Kingdom, which has a general tendency to collect at the surface of the fermenting liquid as attenuation becomes advanced: secondly, the alcoholic ferment in general use in pursuance of the German method, the general tendency of the yeast in this case being to settle at the bottom of the fermenting vessels.   The following terms denoting the difference of behaviour of the two ferments, are in somewhat general acceptance, viz., Surface and Sedimentary, High and Low, Top and Bottom, yeast.

The differences exhibited on a microscopical examination of English and German liquid yeasts are not so strongly marked as might be imagined.   Plate VII., Fig. 1, shows Lager-beer yeast according to two different authorities : it exhibits very much the same rounded and oval forms, of about the same average diameter as the English yeast ; the vacuoles are sometimes more plainly visible than in this last, and there is doubtless a greater tendency for the newly-formed cells to remain attached to the parent cell, owing to the placid nature of the fermentation ; thus causing a more frequent occurrence of cells in pairs, or groups of cells containing a greater number than two. We may say, however, that several samples of Lager-beer yeast that we have examined, show the cells in a fairly dis-associated condition.

A few words in connection with the Continental Lager-beer process will, we think, not be out of place. Considerably less trouble appears to be taken in the production of malt than is the case in this country ; it is grown up less, 8–9 days on the floors being the average time ; and is a much shorter time on the kiln ; malt for Bavarian-beer being dried in from 36–48 hours, whilst a less time suffices for the malt used for Vienna and Bohemian beer. The drying temperatures given by Thausing, for Vienna and Bavarian beer-malt, are somewhat higher than we employ ; but this may be entirely a matter dependent on kiln construction. A very full extract is obtained in the mash-tub on the decoction system ; and the worts are somewhat lightly hopped, but an excellent quality of hop is usually employed, and added to the copper in two portions, the first being boiled from $1\frac{1}{2}$–2 hours, the second $\frac{3}{4}$–1 hour. For the lighter kind of running ales (Schank-bier), about 6 lbs. of hops per quarter of malt are used, whilst for store or Lager-beer 8 or 9 lbs. would be the quantity. The worts— of specific gravity averaging about 1052·5—are cooled to a low temperature—40°–50° F.—according to the class of ale required ; the fermentation lasting from 8–14 days. A very moderate head is formed, appearing first as a slight froth, then taking on a crinkled appearance ; in fact, making allowances for the sluggish nature of the fermentation, the surface changes bear comparison with those we are familiar with in this country. A small amount of yeast is contained in the head, but it eventually settles almost completely.

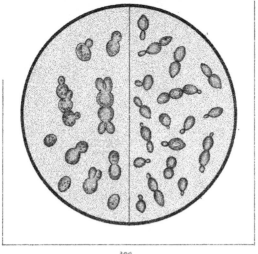

$$\frac{300}{1}$$

Fig 2.

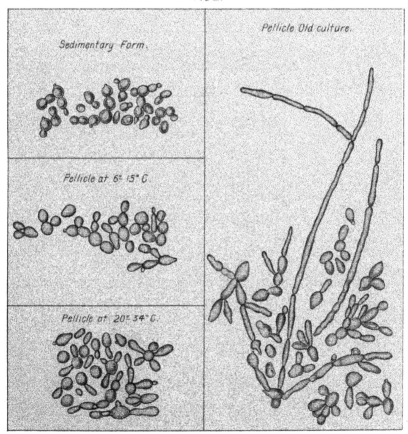

Pellicle Old culture.

Sedimentary Form.

Pellicle at 6°-15° C.

Pellicle at 20°-34° C.

Growth of S. Cerevisiae I (after Hansen)
Reduced from 1000 to 300 diameters

During fermentation, the temperature rises a degree or two. Lager-beer is usually racked into somewhat capacious casks, of varying size, holding on an average some 25 barrels, and stored in a cellar whose temperature is kept as near as possible to the freezing point, for three months.

The process of Pasteurization or sterilization of bottled beer, is somewhat extensively employed in Germany; it consists in submitting the beer in bottle to a temperature of about 130° F. for half-an-hour or so, thus securing practical immunity from change. The process is obviously not applicable to English beers, which require living ferments to ensure the necessary secondary fermentation: whereas the German beers being bottled at a low temperature, contain an amount of Carbonic Acid gas, which on expansion by rise of temperature, ensures the requisite condition.

Low yeast seems to exhibit activity throughout a somewhat wide range of temperature; its action can go on as low as the freezing point, and on the other hand it may be, and is, frequently employed for bakers' purposes, exhibiting a degree of activity in this respect, far superior to most English yeast. As mentioned under Caseous ferments (page 48), we have ourselves obtained the phenomenon of surface fermentation, from a low yeast having a fairly powerful action at 60°—70° F., and capable even of fermenting at 80° F. It constituted also an excellent bakers' yeast.

The question naturally arises from such considerations as these, as to what is the nature of the connection, if any, between "low" and "high" yeast. Pasteur,* after first expressing the view that the different types of Brewing yeast might be modifications derived from some original type of ferment—the existing differences in action having become hereditary—comes eventually to the conclusion

* "Studies on Fermentation," trans. Faulkner and Robb, pp. 187—191, *et seq.*

that "high" yeast is a distinct species; but he herewith proceeds to describe more than one species of high yeast, and was probably at no time working with pure cultivations even of these. Reess* expresses his opinion in the following manner:—Striking as may be the differences between the vegetation of "low" and "high" yeast, it does not allow of their division into distinct species. "Low" yeast can grow and bud at temperatures 9—18° F. higher than those employed in "low" fermentation, but the out-put of yeast is small, and in a single experiment the appearances of high yeast are not arrived at. On the other hand, S. Cerevisiæ of "high" ale-fermentation, set at 40°—43° F., vegetated after six days, in typical "low" yeast forms : hence Reess considers that the "high" and "low" ferments may be modifications of the same species. Our own view in connection with the matter is this :— That where the morphological and chemical functions of different ferments are not very different, and temperature most favourable to action is the chief variable, it is more than probable that by gradual acclimatisation, the ferments could be brought to exercise their action at the same range of temperature. Some recent work of Dr. Dallinger† on Monads, seems to us by analogy to favour this view. During seven years he applied to a certain kind of low organisms termed Monads, a range of temperature commencing at 60° F., and cautiously raised, or held as the occasion required, month by month till 158° F. was attained, the organisms still living and multiplying : a temperature far below this being immediately fatal to unacclimatised organisms. An accident unfortunately terminated the experiment at the temperature last named. It was calculated that at least half-a-million generations must have been produced. From our point

---

* Untersuchungen über die alcoholgährungspilze.—Dr. Max Reess, p. 8.

† Journal Royal Mic. Society, Feb. 9, 1887.

of view then, it seems possible that existing forms in "low" yeast have their specific representatives in "high" yeast, or *vice versâ*, and that the various kinds of Saccharomyces may be modified or educated into carrying on their fermentative action at essentially different ranges of temperature.

The foregoing argumentative matter leads us by a natural gradation, to a consideration of the efforts that have been made to select and cultivate particular species and varieties of Saccharomyces. Hansen has been the chief investigator here, and his work, besides being of much scientific interest, has been productive of practical issues of much importance to the Continental Brewer, and as some think may eventually have its effect upon the English process. Until comparatively recent years Lager-Beer yeast seems to have been of pretty much the same heterogeneous character as British yeast, and like it may be considered to have contained a preponderance of some particular species of yeast which had survived amidst unfavourable conditions, and was best calculated to carry out a fermentation at the low temperatures employed, and hold its own against the foreign ferments present; these last however, would—as in this country—gain the upper hand now and then, and cause serious trouble, such as frets in the finished ale, persistent cloudiness and unpleasant flavours, besides the incursion of Bacteria and, as a possible accompaniment, the complete spoiling of the ales. Now, although Pasteur had more than hinted that foreign or wild yeasts might be a source of trouble, Hansen was one of the first to perceive that practical immunity of yeast and beer from Bacteria did not by any means imply freedom from abnormal secondary fermentation, and to him belongs the credit of having by a succession of ingenious researches, elaborated a practical method for the differentiation and cultivation of species or varieties of yeast that are distinctly favourable in their action, as compared with those kinds that are distinctly the reverse. We purpose somewhat generally reviewing the steps by which Hansen arrived at the present standpoint.

In 1879[*] he published the first communication on the organisms which at different periods of the year are found in the air at Carlsberg and its environs, the said organisms being susceptible of development in Beer wort. The results accruing from this research we have detailed in connection with air (Chap. X.); for our present purposes it is sufficient to remark that a great variety of Saccharomycetes were encountered, with organisms of other classes. Some of these organisms were studied separately, and amongst them Saccharomyces Apiculatus.[†] We have already alluded to the main facts elicited in this research, in the prosecution of which new methods and apparatus were devised for the cultivation of the organism in a state of purity.

In 1882 a further communication[‡] appeared, dealing with the organisms found in the air of Carlsberg and its environs; the method of air testing by exposure of flasks of previously sterilized beer-wort to the local infecting influences, being applied to a determination of the percentage of organisms in different parts of the Old Carlsberg Brewery; this percentage being found to vary very much according to the particular location. Certain occurrences in the Carlsberg and other Danish breweries induced Hansen to experiment with a variety of S. Pastorianus, obtained from some of these air sown cultivations; and on carrying out experimental fermentations of Beer-wort with it, he found that the beer so produced had always a particular odour and a disagreeable bitter taste; brewings carried on side by side with S. Cerevisiæ giving a normal

---

* Meddelelser fra Carlsberg Laboratoriet. Andet Hefte, 1879.

† Meddelelser fra Carlsberg Laboratoriet. Tredie Hefte, 1881.

‡ Resumé du Compte Rendu des travaux du Laboratoire de Carlsberg. 1er vol., 4e livraison.

beer from the same wort. Hansen contrasted this state of things with what had occurred in practice in the Danish breweries, and concluded that this form of S. Pastorianus was a fruitful source of trouble, and as associated with a variety of S. Ellipsoideus was *the* source in the case of the Tuborg and Alt Carlsberg breweries. From the impure yeast of Alt Carlsberg four kinds of Saccharomyces were separated, of which only one, now known as Carlsberg low yeast No. 1, gave a normal beer ; of the other forms, that designated S. Pastorianus I. was the chief cause of mischief.* The value of the results was at once apparent, and in 1884 yeast selection on a practical scale was an accomplished fact. Returning now to some of the detail by which this end was attained. The first thing to be done was to secure pure cultivations from single cells. Nägeli, Lister, Klebs and Koch had paved the way to this, by arriving at an estimate of the number of Bacteria contained in a given portion of an infected liquid, and then diluting it to the extent necessary to give one individual in a definite small volume ; but the difficulty was to ascertain beyond a doubt, that the observed growth following the infection, proceeded from one germ alone. Hansen working with yeast was able to settle this point, from the fact of the cells that were sown, adhering to the walls of the culture flasks and forming a spot or colony as the growth proceeded.

Koch devised a method of dilution and subsequent cultivation in nutrient gelatine, which answered well for both yeast and bacteria, as the colonies formed remained undisturbed, unless they merged into each other, or a liquefaction of the gelatine took place. In the meantime Professor Panum of Copenhagen, had brought into more general application an instrument known as the Hæmatimeter, for counting the organisms in a given area ; Rasmus

---

* Untersuchungen aus der Praxis der Gahrungsindustrie. Dr. Emil. Chr. Hansen. 1 Heft, p. 12.

Pedersen applying the same to the counting of yeast cells more especially.* Hansen subsequently proceeded to devise a modification of existing culture apparatus whereby he could determine whether a growth proceeded from one or several individuals.

We will now proceed to describe processes embracing the foregoing, first mentioning a method that we have employed ourselves for calculating the number of cells present in a given volume of liquid. A drop of ordinary liquid yeast is stirred into 100 cubic centimetres of sterilized distilled water, and a portion measured as follows. An eyepiece micrometer ruled in squares, is fitted to the Microscope, with an objective that magnifies some 40—50 diameters, rendering the yeast cells just visible. The relation of the micrometer scale to a Stage micrometer divided into 100ths and 1000ths inch is ascertained. A square cover glass is taken; its sides measured by the eyepiece micrometer ; and the area calculated. Next the area of the visible field is computed from its diameter as measured by the micrometer scale. The area of the field divided into the area of the cover glass gives the number of fields that could be provided by the cover glass.† A loop say $\frac{1}{8}$ inch diameter, is made at the end of a piece of ordinary platinum wire, and then bent at right angles to the shank, so that when dipped into any liquid, a drop of such size is taken up as will, when placed on a slip and the cover glass put on, fill up all the space under this last. The drop having been previously weighed by hanging the wire to the hook of a chemical balance, we have now all the data necessary to calculate the number of cells in a given volume of liquid, and dilution can be carried out so that 1 cc. contains 1 cell or any desired number : for instance, supposing dilution were first carried out till only

---

* Meddelser fra Carlsberg Laboratoriet. Förste Hefte, 1878.

† This may be preserved after valuation.

one cell per square of the micrometer scale had been first exhibited, the next dilution could be arranged to give one cell per field, and a last dilution, regulated by the number of fields in the cover glass, would give one cell per drop.

To ensure the more complete dispersion of the yeast in water, and to obviate froth, Hansen adds a trace of dilute Sulphuric acid (1 : 10). He also uses a capillary tube to provide a drop of known volume, and the Hæmatimeter for counting; this latter consists of a shallow glass cell, made by cutting a circle out of a cover glass, and then cementing the remainder to a glass slip. The drop fills up the space of the shallow cell when a cover glass is put on. The portion of the glass slip forming the bottom of the shallow cell, may be divided by ruling on some known scale; or the cover glass is ruled. The cell is 0·1 millimetre deep, and the ruled squares usually 0·0025 mm. square. The cubical capacity of each small square then equals .00025 cc. Hansen recommends that 48 to 64 squares be counted, in order to arrive at an average of cells per square, or per cubic centimetre.

A suitable dilution having been obtained, some nutrient liquid or medium may now be infected. First, as an example of a liquid :— Bitter wort, as collected bright from ground bags, affords a suitable nutrient solution for most of the Saccharomycetes, and quantities may be collected in Pasteur Flasks (Fig. 18), which are the most convenient form of apparatus for such work. The flask having been two-thirds filled with wort, is raised to boiling on a sand bath, steam first issuing from the side tube *a*, provided with a piece of caoutchouc tubing, which is then stopped with

Fig. 18.

PASTEUR FLASK.

6

a glass rod. Steam next issues from the long tube, and this after a short interval is in turn closed with an asbestos plug. The sterilization should now be complete, and the contents of the flask may remain for years practically unchanged. As the flask cools, the air entering is filtered through the asbestos, and any germs passing it are deposited on the sides of the tube, which can be re-sterilized at any time by external application of heat.

Fig. 19.   Fig. 20.

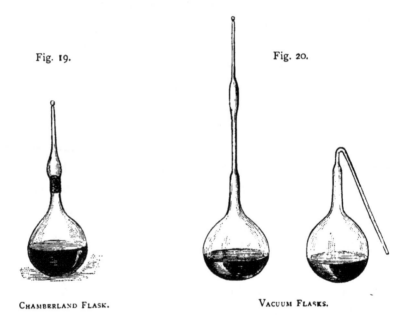

CHAMBERLAND FLASK.   VACUUM FLASKS.

The Chamberland flask (Fig. 19) is a convenient form for some experiments. If it be desired to keep the liquid sterile in vacuo, the forms shown in Fig. 20 may be employed; the tube being plugged with wool or asbestos can be bent over; and sealed off by a suitable flame, such as the blow-pipe, during the drawing in of air after boiling. The simplest form of culture flask is the conical one (Fig. 21), in which the liquid is sterilized in the ordinary way, and a piece of sterilized filter-paper is secured over the mouth by an india-rubber ring or other means.

The flasks are infected with the desired organisms by

introducing the necessary portions of liquid rapidly, with all precautions requisite to ensure sterility of implements used, and in a room as free as possible from floating dust. In the flask Fig. 18, the introduction is made through the short tube, and the stopper immediately replaced. In Fig. 19, the ground neck and cap are conveniently smeared with a little vaseline before sterilizing the liquid: the cap being removed the infecting liquid is introduced. In the

Fig. 21.

Assay Flask.

case of Fig. 20, the tube is allowed to dip into the infecting liquid, and the point being broken off under it, the vacuum causes an in-rush of liquid which may be controlled as desired. With Fig. 21, it is simply a matter of taking off and replacing the old or a fresh paper covering. In each case the flask after inoculation, is submitted to a favourable temperature in an incubator or warm chamber. Where yeast is sown, it falls to the bottom of the flask and fermentation starts, the points of growth being noted : each speck appearing indicates a colony. Where the object is to procure only one colony in a flask, it is usually desirable to set a certain number with, say 1 cc. of a liquid containing an average of $\frac{1}{2}$ a cell, that is, 1 cell to every 2 cc. Let us suppose the colony resulting from the growth of a single cell to have been obtained in a Pasteur flask, and that a larger quantity of pure yeast from the same "store" is required. Most of the beer is run off through the small side tube ; the remainder is shaken up with the yeast ; and the whole removed by connecting the short tube by a caoutchouc tube, with the orifice of some large tinned copper vessel holding some gallons of wort, which has been sterilized in it by boiling and subsequently cooled. Here sufficient yeast is produced to set a much larger fermentation, and so eventually enough is obtained to pitch a square. For further detail we must refer our readers to Hansen's own communications,* or translations and abstracts of them in the Brewing Journals.

Hansen adopted at a certain stage in his experiments, a solid medium for the initial cultivation of the selected cells, consisting of a mixture of hopped wort and gelatine. It was thus made possible to trace the development of the single cells under the microscope. The method is as follows :— A specimen of the yeast is largely diluted with water in a Chamberland flask ; drops of this are further diluted with hopped wort and gelatine—wort of Sp. gr. 1058, and 5—10°/₀ of gelatine, filtered bright, or fined and filtered—

Fig. 22.

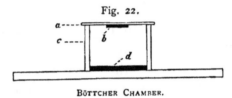

BÖTTCHER CHAMBER.

contained in test-tubes in which it has been sterilized and cooled to 70—75° F. Complete mixture is effected, and a drop of the gelatine wort, now containing yeast, is examined microscopically, and should show only a cell or two to a field. Dilution to any desired extent may now be carried out ; a drop being finally withdrawn by a sterilized glass rod, and spread on the under surface of a thin cover-glass, which is placed on the ring of a Böttcher or Ranvier moist chamber.† [Figs. 22 and 23.]

The Böttcher chamber [Fig. 22] consists of a glass cell formed by cementing a ring (*c*) to a slide ; water is placed in the bottom (*d*), and the position of the drop of gelatine is at *b*, on the under side of the cover-glass *a*. The

* Untersuchungen aus der Praxis der Gahrungsindustrie.—Dr. Emil Chr. Hansen.

† These moist chambers appear to us to be a modification of Van Tieghem and Lemonnier's cell for examination of moulds.
Compare "Etudes sur la Bière," Pasteur, p. 153 ; trans. p. 155.

Ranvier chamber (Fig. 23) is a modification of the fore-going, the water receptacle being an annular groove *a, a,* ground out from a slip, the portion enclosed *c* having its height reduced to afford space for the drop of gelatine enclosed between the cover-glass *b,* and *c;* the edges of the cover-glass projecting beyond the circular groove. Vaseline may be smeared on the surfaces which come in contact, so as to secure air-tight connections. In either case one or two cells are picked out, and their position marked by a diamond on the cover-glass; the apparatus

Fig. 23.

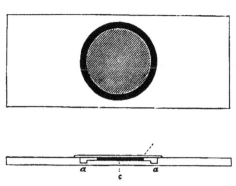

RANVIER CHAMBER.

is then put in an incubator at 80°—90° F., and left for a day or two before re-examination. The specks of yeast may be taken up by a short piece of Platinum wire, and the wire dropped into sterilized hopped wort of Sp. g. 1058, contained in Pasteur flasks.* The growth being known to proceed from a single cell, the required conditions are fulfilled. Through the whole of Hansen's work there is a pervading idea that the shape, size, and appearance of the cells did not in themselves suffice to confer a distinct individuality as regards species or variety; for one and the same kind of Saccharomyces was found capable of exhibit-ing a variety of forms, corresponding to changes in the

* Résumé du Compte Rendu des Travaux du Laboratoire de Carlsberg. 2me vol., 4me liv., 1886.

conditions attending development. As an example: a "low" yeast growing with difficulty at 80° F. may give long branching cells, whilst at 45°—50° F. it would give the well-known low yeast forms. On this was based a test that Hansen employed for low yeast (and by which he identified several species or varieties that had specific actions in the Brewing process, that we shall speak of later on), namely, to first examine microscopically the sedimentary form produced at normal temperature, and then to set the yeast afresh, and allow it to develop in wort at 78°—80° F. These purely microscopical investigations were followed by observation of peculiarities in the mode of growth, and notably the conditions under which Ascospores were formed ;* and from the variations shown in this latter respect under like conditions of temperature, etc., Hansen collected further means of identification of the differentiated ferments—

S. Cerevisiæ I.

S. ·Pàstorianus I.

,, II.

,, III.

S. Ellipsoideus I.

,, II.

We have already in Chap. III., spoken of the peculiar mode of reproduction by ascospores, and the way in which the phenomenon is obtained. Hansen employed sterilized blocks of plaster of Paris, on which the yeast was poured, and which were then put into shallow glass dishes half filled with water, and covered up. For ordinary ascospore formation a temperature of 60° F. suffices. It was found that sporulation proceeded very slowly at low temperatures, becoming more rapid as the temperature rose, till a point was reached where the development was again restrained, and finally ceased entirely. The lowest temperature for the

---

* Résumé du Compte Rendu des Travaux du Laboratoire de Carlsberg. 2me vol., 2me liv., 1883.

six kinds of yeast first treated was 33° F. or ·5° C.; the highest limit, 99·5° F. or 37·5° C. Between these limits there were characteristic landmarks for the different kinds of yeast, which enabled a separation to be made. The results are concisely and conveniently expressed in the following table, taken from a paper read by Dr. G. H. Morris in 1887 [*] The equivalents of the Centigrade temperatures in Fahrenheit degrees have been added.

ASCOSPORE FORMATION.

Tempera- ture. Faht.	Tem- perature. C.	S. Cerev. I. (Hansen.)	S. Past. I. (Hansen.)	S. Past. II. (Hansen.)	S. Past. III. (Hansen.)	S. Ellips. I. (Hansen.)	S. Ellips. II. (Hansen.)
99·5	37·5 C.	None.	—	—	—	—	—
96·8—98·6	36—37	29 hours.	—	—	—	—	—
95	35	25 ,,	—	—	—	—	None.
92·3	33·5	23 ,,	—	—	—	None.	31 hours.
88·7	31·5	—	None.	—	—	36 hours.	23 ,,
86	30	20 hours.	30 hours.	—	—	—	—
84·2	29	—	27 ,,	None.	None.	23 hours.	22 hours.
81·5	27·5	—	24 ,,	34 hours.	35 hours.	—	—
79·7	26·5	—	—	—	30 ,	—	—
77	25	23	—	25 hours.	28 ,,	21 hours.	27 hours.
73·4	23	27	26 hours.	27 ,,	—	—	—
71·6	22	—	—	—	29 hours.	—	—
64·4	18	50	35 hours.	36 hours.	44 ,,	33 hours.	42 hours.
61·7	16·5	65	—	—	53 ,	—	—
59	15	—	50 hours.	48 hours.	—	45 hours.	—
51·8—53·6	11—12	10 days.	—	77 ,	—	—	5·5 days.
50	10	—	89 hours.	—	7 days.	4·5 days.	—
47·3	8·5	None.	5 days.	—	0 ..	—	9 days.
44·6	7	—	7 ,,	7 days.	—	11 days.	—
37·4—39·2	3—4	—	14 ,,	17 ,,	None.	None.	None.
32·9	0—5	—	None.	None.	—	—	—

* "The pure cultivation of Micro-organisms, with special reference to yeast."— *Journal of the Society of Chemical Industry*, Feb. 28th, 1887.

It will be noticed that the maximum and minimum temperatures for the different species, are in themselves different; as also the limits of temperature within which the ascospore formation takes place in the species examined. The differences are the most marked at the lower temperatures.

Holm and Poulsen[*] using the foregoing methods have succeeded in detecting with certainty the presence of $\frac{1}{200}$ of wild yeast in a sample of pitching yeast; this is of the more interest from the fact that Hansen had previously found that a mixture of S. Pastorianus III. and S. Ellipsoideus II.—which present in quantity may cause turbidity in beer—does not do so when in the proportion of $\frac{1}{11}$ of the pitching yeast. The method possesses the advantage of being moderately quick, for in the above-mentioned cases a cultivation at 25° C. (77° F.) would give a plain indication after 40 hours; the wild yeasts sporulating very soon.

The problem of differentiating Brewery yeast has been further attacked by Hansen—apparently with success—by noting the conditions under which films or pellicles of Saccharomyces are formed on the surface of fermentable liquids; the said films or pellicles appearing to us to be pretty much the same thing as Pasteur's aerobian ferments.[†] This phenomenon is, as we know, not confined to Saccharomyces, but the films of the alcoholic ferments differ in appearance from those of Mycoderma Vini, Bacterium Aceti, etc. The films form generally towards the close of flask fermentations; small islets of yeast being carried to the surface collect, and develop a greyish yellow slimy film, which if partly shaken down renews itself. A free, quiet liquid surface, with direct entry of filtered air,

---

[*] Résumé de Compte-Rendu des Travaux du Laboratoire de Carlsberg. 2me Vol., 4me Liv. 1886.

[†] "Studies on Fermentation," trans. by Faulkner and Robb, p. 205, *et seq.*

is essential to the growth. The assay flask covered with a filter paper answers well. During the growth the colour of the wort becomes reduced to a light yellow.

The points Hansen set himself to determine were :—

1. Temperature at which the film formed.

2. Time elapsing before the appearance of the film at different temperatures.

3. Microscopical appearance at different temperatures. Cells of old films show striking varieties of form. Young films of S. Cerev. I., S. Past. II., and S. Ellips. II., show no mycelial or branching colonies; these last are, on the contrary, found with S. Pastorianus I. and III. and S. Ellips. I.

At high temperatures only S. Cerev. I. and S. Ellips. II. appear to vary from the others, but at 13°—15° C. (55·4°—59° F.), with young films, it is very different; for instance, S. Past. II. and III., which are upper or high ferments—whose cells in ordinary sowings look the same—give quite a different vegetation; and an equally striking difference obtains between S. Ellips. I. and II. · at this temperature then it is only a matter of difficulty to distinguish between S. Past. I. and II., and here the ascospore formation helps, as also the circumstance that in flask fermentation at ordinary temperatures, one is a "high" and the other a " low " yeast.

As a typical example of these pellicle growths we give on Plate VII., Fig. 2, the appearances exhibited by S. Cerevisiæ I., a ferment separated from a sample of Edinburgh yeast. A reduction to ordinary magnification has been made for purposes of general convenience, and to permit of comparison with Fig. 1 representing Lager-Beer yeast.

The following table, taken from Dr. J. H. Morris's paper* already quoted, gives the facts connected with the film formation in a comprehensive manner.

* J. Soc. Chem. Ind. Feb. 28, 1887.

FILM FORMATION.

Temperature. Fahᵗ.	Temperature. Cenᵗ.	S. Cerev. I. (Hansen.)	S. Past. I. (Hansen.)	S. Past. II. (Hansen.)	S. Past. III. (Hansen.)	S. Ellips. I. (Hansen.)	S. Ellips. II. (Hansen.)
104°	40° C.	—	—	—	—	—	None.
96·8—100·4	36—38	None.	—	—	—	None.	8–12 dys.
91·4—93·2	33—34	9–18 dys.	None.	None.	None.	8–12 dys.	3–4 ,,
78·8—82·4	26—28	7–11 ,,	7–10 dys.	7–10 dys.	7–10 dys.	9–16 ,,	4–5 ,,
68—71·6	20—22	7–10 ,,	8–15 ,,	8–15 ,,	9–12 ,,	10–17 ,,	4–6 ,,
55·4—59	13—15	15–30 ,,	15–30 ,,	10–25 ,,	10–20 ,,	15–30 ,,	8–10 ,,
42·8—44·6	6—7	2–3 mths.	1–2 mths.	1–2 mths.	1–2 mths.	2–3 mths.	1–2 mths.
37·4—41	3—5	None.	5–6 ,,	5–6 ,,	5–6 ,,	None.	5–6 ,,
35·6—37·4	2—3	—	None.	None.	None.	—	None.

In the above table, dys. = days ; mths. = months.

The chief properties of the ferments that have been spoken of in connection with ascospore and film formation, are given in the following summary.[*]

S. Cerevisiæ I.—A "high" yeast obtained from the pitching yeast of an Edinburgh Brewery, and afterwards from that of a London Brewery, develops ascospores at temperatures between 11° C. and 37° C. ; the greater number of the cells resembling the original yeast ; film formation at 13°—15° C.

S. Pastorianus I.—Obtained from air-dust in the neighbourhood of the Carlsberg Brewery, Copenhagen, is a "bottom" ferment resembling Pasteur's.[†] It imparts a bitter flavour to beer, develops ascospores at temperatures between 3° C. and 30.5° C. ; film formation at 13°—15° C.

S. Pastorianus II.—From air-dust. Rather larger than Pasteur's or Reess's form. Acts mainly as a bottom ferment. Causes no disease in beer. Develops ascospores between 3° and 28° C. ; film formation at 13°—15° C.

S. Pastorianus III.—From a low-fermentation beer

---

[*] Derived mainly from "Micro-organismen der Gährungsindustrie" (A. Jörgensen).
[†] Etudes sur la Bière. Trans. "Studies on Fermentation," Faulkner and Robb, Plate XI.

produced in Copenhagen, attacked by yeast turbidity (S. Ellips. II. present at the same time). Develops ascospores between 8.5° and 28° C. ; film formation at 13°—15° C.

S. Ellipsoideus I.—Obtained from the surface of ripe grapes from the Vosges. Effects on beer not yet investigated. Resembles Pasteur's and Reess's form. Develops ascospores between 7.5° and 31.5° C. ; film formation at 13°—15° C.

S. Ellipsoideus II.—Associated with yeast turbidity in beer. Develops ascospores at temperatures between 8° and 34° C. Film formation at 13°—15° C. Resembles the ordinary form, S. Ellips. I., in a marked degree.

With the exception of S. Cerevisiæ I., which constituted the larger proportion in some samples of English yeast, the above-mentioned forms are to be regarded as wild yeasts, and as contaminating influences in relation to normal healthy yeast, for their presence in appreciable quantity may render an otherwise normal brewery yeast incapable of producing a beer of good flavour and keeping properties. Contamination with wild yeasts may be produced by the dust of the air during summer and autumn, and may originate from other sources.

And now as to some of the practical results of yeast selection. In the first place, two varieties of S. Cerevisiæ have been separated from " low " yeast, and are employed in the Carlsberg and other breweries, being known respectively as No. 1 and No. 2 Carlsberg yeasts.

No. 1—obtained from the yeast which had been in use in the Carlsberg Brewery for many years, and which was originally brought from Munich—gives a beer of much stability, somewhat thin on the palate, and containing less Carbonic Acid gas than that from No. 2. The beer clears very slowly in the Lager cellar, but when bright, should remain so for at least three weeks after bottling, for which last it is well adapted.

No. 2—isolated from Hamburg yeast, and later from that of other places—gives a nice flavoured beer of much palate fulness, which contains more Carbonic Acid gas than No. 1. The Beer is not adapted for bottling, having keeping properties inferior to that from No. 1.

As an indication of the scale on which pure cultivation has been carried out, it may be mentioned that Hansen, starting from a single cell, produced in a few weeks the whole bulk of yeast (5,500 lbs.) employed in the Alt Carlsberg Brewery. For some four or five years selected yeast has been in continual use in the Carlsberg Breweries, being periodically produced by an inter-dependent process. The two Carlsberg Breweries together turn out some 400,000 hectolities of beer (say 244,000 Barrels) per annum. Selected yeast is used besides, in some of the breweries of other countries, and seems to give satisfaction. In addition, some few breweries are working with yeast selected by Hansen's method from their own original pitching yeast.

Under favourable conditions, the pure selected yeast does not soon deteriorate ; for instance, in Alt Carlsberg the No. 1 yeast has been kept pure for 6 to 8 months, and the No. 2 yeast, 2 to 4 months, even with free exposure to air of the worts on the cooler.

Where yeast is to be examined for wild forms, Hansen recommends that the samples be taken at the end of the primary fermentation : if for the selection of cells of the normal ferment, the samples should be taken at the commencement of fermentation.

A point of some interest in connection with Hansen's work, is the evidence he adduces as to the persistency of form, in the progeny of differently shaped cells of one and the same species or variety of yeast ; spherical cells producing spherical cells, and oval cells those of an oval form ; though, with long continued cultivation, a tendency to produce one type of cell becomes more and more marked, till finally, all the cells approximate to the same shape.

Certain species of Saccharomyces, namely, S. Exiguus, S. Minor, and Reess's S. Conglomeratus—the latter of which does not seem to be associated at all with beers—have not as yet been put to the test of pure cultivation.

In addition to the Alcoholic ferments that have been already mentioned in this and the foregoing chapter, there remain a few of rarer occurrence, of which we need only mention in the briefest manner.

Saccharomyces Marxianus (Hansen).—Forms small cells like S. Ellipsoideus and S. Exiguus, also very irregular forms. Does not yield ascospores readily. Gives only a feeble fermentation in beer-wort, being unable to ferment Maltose.

Saccharomyces Membranæfaciens (Hansen).—Produces a clear grey pellicle on beer-wort. Forms oval or elongated empty-looking cells. Yields spores abundantly. It does not ferment beer-wort. Appears to be inactive with most sugars ; and is unable to invert Cane Sugar. It liquefies gelatine very readily. Hansen appears to class it as a Saccharomyces on account of the spore formation, but, in face of the evidence he adduces, it would seem to be rather straining the title.

Excellent as is the whole of Hansen's work, and of unquestionable physiological interest and importance, it still remains an open question as to whether a degree of purity of Lager-beer yeast consistent with the requirements of practice, could not have been secured by other means than single-cell selection ; such for instance, as attention to the plasmatic conditions most favourable to the action of one species of yeast. M. Velten, of Marseilles, one of Hansen's chief antagonists, evidently has views of this kind in connection with the matter. He gives the preference to what may be called Pasteur's normal yeast, that is, a bacteria-free yeast containing a preponderance of a desired species, and whose foreign or wild yeast forms are essential to a proper secondary fermentation. Our own experience of the persistence of certain types of yeast in this kingdom, such as London, Edinburgh, Yorkshire Stone Square, and Burton yeasts, leads us to attach no little importance to Velten's views ; and it seems to us that the case may be stated as follows :—From Velten's point of view the conditions of the process should be adapted to secure the production of a yeast of uniform type, starting presumably, with a yeast that has before given satisfactory results. Hansen's contention is practically the converse of this—viz., that a yeast must be selected to suit the process.

Hansen himself admits that the pure selected yeast will not do everything, and putting aside for the moment any consideration of its adaptability to English beers, there are some cases where pure yeast would manifestly not produce the required result. We speak of some of the Belgian breweries in which beers are still produced by spontaneous fermentation.

The methods of brewing pursued in the United Kingdom are certainly such as to encourage the development of more than one kind of Saccharomyces, and we know that our beers are open to defects similar to those that Hansen encountered. Granting that a selection could be made of a typical cell suited to each mode of fermentation carried out in this country, and which would give the desired flavour and normal secondary fermentation, the present conditions of the process would, we think, necessitate a frequent production *de novo* of the typical yeast to replace the degenerated ferment. Experiments on an industrial scale with different species of pure yeast have been carried out at Burton-on-Trent by H. T. Brown and

G. H. Morris, and it is their expressed opinion that many difficulties have yet to be surmounted before the English Brewer can place the same reliance on pure cultivated yeast as his Continental *confrère* is able to do, but that, when these difficulties have been surmounted, pure yeast culture in a more or less modified form, will play a very important part in our English Breweries.*

* H. T. Brown's Introduction to "The Micro-organism of Fermentation." A. Jörgensen. Edited from the German by G. H. Morris.
ᛏ G. H. Morris, Soc. Chem. Ind., 1887, p. 122, and Brewing Trade Review, 1888, page 387. "Alcoholic Ferments, etc."

CHAPTER VI.

The Moulds or Microscopic Fungi.

OWING to the great variety of form exhibited by Fungi generally, varying as they do from the huge toadstool and mushroom, to the minutest mould, any complete classification of even the moulds alone, including as it would such a vast number of species, would be too complex a matter to be either interesting or useful from the Brewing point of view.

The term Mucorini has been applied by Nägeli somewhat generally to the Microscopic Fungi, but it is more often used in classification of the moulds, to denote a certain group with a habit of growth similar to Mucor Racemosus. The term Mucedines has also been used somewhat in the same way, but correctly it has even a more limited meaning.

In comparison with the Saccharomycetes and as we shall see, with Bacteria, the Moulds occupy a position of secondary importance as associated with Malting and Brewing; still, as having at times a definite influence on the process, we are warranted in giving this class of organisms something more than passing notice, and so we purpose describing them in general terms, and briefly to sketch the mode of growth of the chief Moulds encountered by the Maltster and Brewer.

Scientifically—although containing none of the green colouring matter, Chlorophyll, common to vegetables—the Moulds are regarded as belonging more especially to the Vegetable Kingdom, occupying a position between Saccharomycetes (the ferments) and Schizomycetes (Bacteria), but showing a close relationship to these two families at either end of their scale, *e.g.*, the ferment forms of certain moulds connecting these last with the Saccharomycetes proper ; the living spores (Zoospores) of other moulds being closely allied to Bacteria.

The appearance of Mould on various objects, such as old boots, stale provisions, etc., etc., is of course a very familiar occurrence, and it is astonishing what very different objects seem able to support moulds of some kind. The Maltster is not unused to its appearance as a bluish-green growth on germinating Barley ; and other rarer moulds may occasionally be noted on the same medium, distinguished by a black or red colour. The Brewer consequently has at times to deal with Malt whose quality has been reduced by mould, and he may occasionally find some of his Hops deteriorated by a like cause ; or again, for want of proper care the wooden vessels of a Brewery may fall into a mouldy state ; and in the best conducted Breweries a certain percentage of the cask plant is open to adverse influences from the same cause.

Let us now use the Microscope for a preliminary examination of some mould growth : this is a case where the examination of the selected mass may be first carried out under a low magnifying power, say 30 or 40 Diameters, so as to give one a good general idea of the growth, and also to help one to dissect out portions for examination under higher powers, say 200 to 300 Diameters. Growths of mould, illuminated by the Bull's-eye condenser are not unfrequently objects of great beauty when viewed through the Microscope, as they may

resemble miniature forests of peculiar and luxuriant vegetation ; marvellous networks enclosing brilliant spheres ; or
miniature hills with snow wreaths on their summits and
slopes.   If then a specimen of Mould be taken from one of
the sources previously indicated, preferably from some
moist mass, it will be probably found to exhibit the
following characteristics :—The portion nearest to the
substance on which the Mould has been thriving, is seen
to consist of interlaced threads or filaments forming what
is called a Mycelium.   On breaking this up with the
end of a glass rod or a needle, the filaments or Hyphæ are
seen to be tubes, and at certain points in these tubes, thin
dividing walls are perceptible, which are termed Septa.
On examining some of the upper portions of the growth,
it is very probable that numbers of small spherical or
oval bodies will be detected, and—if care be exercised
in the manipulation—will be found to occupy the position in which they were originally formed ; these are
Spores, capable in most cases of giving rise to a fresh
growth of the mould, when falling into a suitable nourishing
stratum ; and it is by the formation of these spores that the
reproduction of the generality of moulds is provided for.
The simplest mode of reproduction witnessed amongst the
moulds, is by the continuous budding and dividing off of
portions of the hyphæ, a process more allied to the true
budding of the Saccharomycetes than to mere fission.   The
next higher mode is the formation of definite naked spores
on the ends of hyphæ, from which they are very easily
detached at maturity.   A stage above this last phase
of growth is the production of spores compacted into a
receptacle (Sporangium or Ascus), the walls of which must
be ruptured before the contained spores can escape.

In the case of some moulds, the spores when ripe issue
from a well defined spore-case, and show active movements
caused by vibrating hair-like appendages called Cilia or

Flagella. Such spores are termed Zoospores or Swarm-spores. The motion continues for a time, and then the spores settle down and germinate.

A curious mode of fructification which appears to be of a sexual order, is not unfrequently exhibited amongst moulds. Two of the threads or hyphæ approach and join each other, and this conjugation is followed by the development of a large spore (Zygospore), capable of giving rise to the mould-growth afresh.

A remarkable phenomenon occurring in connection with Mould growth is what is called Alternation of Generation or Polymorphism ; that is to say, a mould may not always follow one mode of development retaining its characteristic appearance, but may instead, go through a cycle of changes, at certain points in which—were the development not traced *ab initio*—one would believe that a perfectly different species was being viewed. A good example of this is furnished by the " red-rust " of cereals, termed Puccinia Graminis, which on a different host—the Berberis—gives rise to what was formerly regarded as a distinct growth, and named Aecidium Berberis. It was found that cereals in the neighbourhood of shrubs of Berberis were generally attacked by rust. By various observations and experiments the identity of these two dissimilar forms has been placed well nigh beyond dispute. We shall have occasion to make a further allusion to this polymorphism in the case of another mould.

Now as regards the conditions which favour the production of mould :—Although of higher organisation than the Saccharomycetes, the moulds seem able to subsist on less complex forms of nourishment. Solutions of mineral compounds, such as the Sulphates of Magnesium, Copper and Zinc, containing mere traces of impurities, will occasionally furnish growths of this class of organism ; whilst substances like Ammonium Acetate and Tartrate, especially if they be somewhat acid, will also afford adequate nourishment. In the case of liquids that are capable of sustaining Saccharomycetes, these last may develop, followed by Bacteria ; and finally Moulds may thrive in the acid liquid so produced, especially with free exposure to air.

In fruit preserves and syrupy liquids, where the percentage of sugar is too high to support Alcoholic ferments, and the amount of Nitrogen perhaps too low for Bacteria, Moulds may grow unrestrained ; though they commonly follow Saccharomycetes and Bacteria, owing to their property of growing in acid liquids, especially fruit juices ; but here again, the acidity may in some cases, be so great as to prevent Alcoholic ferments or Bacteria growing, whilst certain moulds would be quite at home under the circumstances.

The chief components of Moulds appear to be of the same nature as those of Saccharomycetes ; the enveloping membranes consisting of a Carbohydrate resembling Cellulose and possessing some degree of toughness and durability ; whilst the contents of the hyphæ and spores are viscid protoplasm.

The large Fungi (Mushrooms, etc.) seem to have somewhat the same chemical composition as yeast.

Very little is known of the substances produced by the growth of moulds in different media. Whatever they may be chemically, there is usually some taste or smell resulting that is objectionable, especially the characteristic mouldy or musty flavour and smell that so many moulds are capable of producing ; and even in the few cases that we shall particularise, where moulds grow in the manner described, the taste of the resulting fluid is generally peculiar if not distinctly unpleasant.

Pasteur showed that certain fungoid growths which vegetated by using the Oxygen of the air, and which derive from oxidation the heat that they require to enable them to perform the acts necessary to their nutrition, may continue to live, though with difficulty, in the absence of Oxygen : in such cases the forms of their mycelial or sporic vegetation undergo a change, the plant at the same time evincing a decided tendency to act as an alcoholic ferment.

The only distinct industrial purposes to which mould-growths are applied, are in connection with this power of forming Alcoholic ferments. It appears besides, that moulds may give rise to a species of Diastase, as for instance, in the preparation of the Japanese " Koji," made from steamed rice on which a yellow dust—the spores of a fungus—is placed, and subsequently allowed to vegetate. " Koji " is capable of liquefying gelatinized starch, and setting up a fermentation in it, giving rise to a kind of Beer—-the Japanese " Saké." Koji is also used in breadmaking and as a source of " Soy." The mould giving rise to these spores is called Eurotium Oryzæ.

We purpose now dealing with some selected varieties of Moulds which are associated with the materials used in the production of Beer, and to a certain extent, with the beverage itself; also with one or two moulds that have a connection with wine. We will take them in order of complexity, beginning with the simplest.

OIDIUM LACTIS is a mould frequently found on the surface of milk, but which grows in nearly all substances that sustain mould-growths generally. It occurs occasionally on crushed germinating Barley; on "grains ;" and notably on pressed yeast, especially German or foreign baker's-yeast.

Hansen found it in sterilized worts that had been infected by germs from the air in the neighbourhood of the Carlsberg brewery; but he states that Beer and wort not directly sown with this mould are little liable to its incursion. The same worker's more recent researches prove, in contradistinction to other authorities, a uniform mode of growth on different plasma, and a rapid development on a suitable substratum with a favourable temperature. Its appearance on a liquid is somewhat like that of Mycoderma Vini, but it is more felted, and whiter-looking. Its mode of growth is of the simplest kind met with amongst moulds. Under the Microscope the snow-white downy coating is seen to consist of a mycelium, the interlacing threads of which are divided by septa into varying lengths ; the pieces so marked off on certain hyphæ differentiating into still smaller pieces, which fall away from each other and reproduce similar lengths and chains of cells ; the smallest cells constituting the nearest approach to spores. (See Plate VIII., Fig. 1.) The elongation of the hyphæ and differentiation by septa seem to go on simultaneously, thus resembling the development of Bacteria. With deficient nourishment there is more tendency to form definite spores or conidia, which reproduce by germination. Submerged in Beer-wort, it appears to be very sluggish in its action. Hansen gives 30° C. (86° F.) as the most suitable temperature for the growth of this mould.

CHALARA MYCODERMA is a mould somewhat resembling the foregoing in its mode of growth, but with a tendency to form spherical protuberances in the elongated cells. Hansen obtained this also from the air in the neighbourhood of the Carlsberg Brewery.

OIDIUM LUPULI is an excellent example of a mould resembling Oidium lactis in its mode of growth ; it is occasionally met with on spent hops, on which it forms a reddish-yellow or salmon-coloured dust, which on microscopical examination, is found to consist of branching cells, merging like Mucor Racemosus into spherical cells, some of which have all the appearance of budding. Many

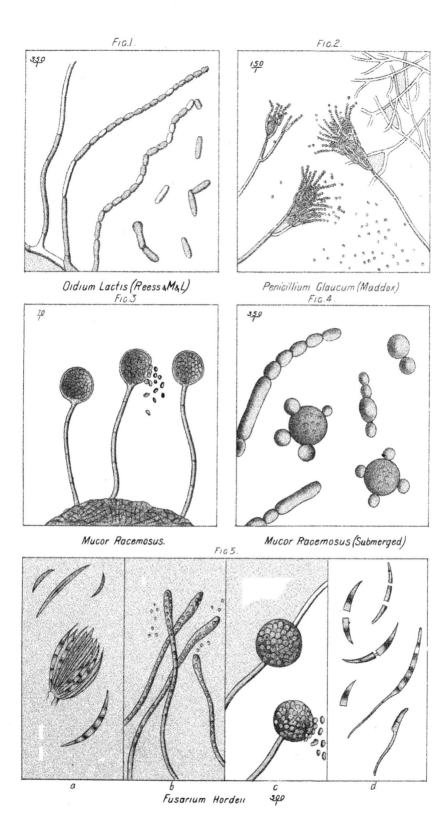

Fig.1.

$\frac{350}{1}$

Oidium Lactis (Reess & M&L)

Fig.2.

$\frac{150}{1}$

Penicillium Glaucum (Maddox)

Fig.3

$\frac{70}{1}$

Mucor Racemosus.

Fig.4.

$\frac{350}{1}$

Mucor Racemosus (Submerged)

Fig 5.

a   b   c   d

Fusarium Hordeii

$\frac{300}{1}$

of the spherical cells and the branching pieces display an orange-pink colour, which seems to permeate the protoplasm. For a purely aerobian form, the mode of reproduction is decidedly interesting.

OIDIUM VINI, also called Erysiphe Tuckeri, is of some little interest, not only from the destruction that it has caused to the French vines in the last 30 or 40 years, but chiefly because it seems to have a specific action in the wine itself, growing submerged, and producing a class of peculiar flavours that render even some of the best wines undrinkable; a marked acidity is also a common accompaniment of its growth. On the Vine, the mould grows in the hyphæal condition, with heads of agglomerated spores. When occurring in wine its appearance is that of hyphæ broken-up into short cylindrical, curved, or branched pieces.

The best method of preventing wines developing this and other mould growths, is to sterilize them by Appert's process, which consists of subjecting the wine for a short time in closed vessels to a temperature above the boiling point of water.

PENICILLIUM GLAUCUM, the most widely spread of ordinary moulds, is distinguished by its bluish-green colour. It appears on fruit, food, etc., but to us, its most interesting occurrence is on germinating barley. It has its origin as a rule, on the half corns and accidentally crushed ones; spreading rapidly under favourable circumstances to sound corns. Its growth appears to be favoured mainly by high temperatures on the malting floors; a large percentage of split or damaged corns; and a decrease in vitality of the germinating barley, owing chiefly to unfavourable atmospheric conditions.

On examining a mouldy corn under successive powers of the microscope, a white mycelium is seen on the surface, from which spring hyphæ or threads, bearing tassels of spores or conidia [Plate VIII., Fig. 2]. These spores are

Fig. 3, exhibits this mould as it occurred on some damp barley contained in a bottle. M. Racemosus is of no little interest in connection with fermentation, as it gives on submersion in a fermentable liquid, a very well-defined ferment form, beginning with huge branching cells which run on into spheres, some of which may be from two to four times the diameter of a yeast cell, and capable of budding at several points, showing in fact, some of the appearances of S. Cerevisiæ ; only on a much larger scale [Plate VIII., Fig. 4]. Its growth is accompanied by the production of Alcohol and Carbonic Acid gas to a limited degree, the maximum amount of Alcohol formed, being according to Fitz, 3·5 to 4% by volume : (an experiment of our own gave us 4% in a wort of Sp. Gr. 1063). A fair supply of Oxygen facilitates its growth.

The mycelium of Mucor Racemosus, taken out of a fermented liquid and exposed to the air on a nourishing medium such as crushed germinating barley, reproduces the aerial or mould form.

Mucor seldom appears in Beer, probably because the fermentative activity of S. Cerevisiæ, being so much greater, a proportion of Alcohol is soon arrived at that precludes its growth. We have heard of its being seen in yeast, but only in connection with plant and process of the most defective description.

Mucor mucedo is also a commonly-occurring mould on rotten fruits, mouldy bread, old yeast, damp barley, and malt ; but it grows perhaps more readily on horse manure than on anything else. It has a passing interest for us, in that its mode of growth somewhat resembles that of Mucor Racemosus, forming as it does, a definite sporangium, at the extremity of hyphæ proceeding from a white or greyish mycelium, usually dark-coloured and visible to the naked eye. Pasteur distinguished it from M. Racemosus by the circumstance of its having on its sporangia-bearing hyphæ, lateral branches which also terminate in Sporangia.

In Beer wort, instead of forming the well-defined ferment forms that M. Racemosus does, it has a greater tendency to form a branching mass of large and long cells, with here and there huge swellings, filled generally with granulated protoplasm and nuclei. It produces small quantities of alcohol.

Any living portion of the original mycelium seems capable of growing submerged in Beer-wort, which is probably also the case with M. Racemosus.

Pasteur, on examining the adherent dust of grapes, found, besides specific Alcoholic ferments, certain forms of moulds, including Dematium Pullulans, which were capable of producing in the grape-juice, cells closely resembling the Saccharomycetes ; there is no evidence however, that he obtained the phenomena of fermentation from the bodies in question.

Brefeld has shown that many of the moulds, cultivated in nutrient liquids, are transformable into *torula* forms or cells resembling yeast, not usually classed as Saccharomycetes ; and he is strongly of opinion that the Alcoholic ferments are the submerged sporular forms or conidial fruit of moulds or fungi.

Black Mould of Hops and Barley.

In certain seasons, Hop and Barley samples are met with that exhibit minute black spots or patches: if these be scraped off with the point of a penknife, and placed with a little moistening liquid—preferably dilute alcohol or dilute glycerine—on a slide and examined with a combination of about 300 diam[rs.] certain mould structures may be observed, which from either source are generally identical, and are characteristic of one of the varieties of Ustilago, probably U. Carbo or U. Segetum, the "smut" of cereals. The chief peculiarity is the dark-coloured (brownish black) hyphæal growth, somewhat resembling Oidium Lactis in form ; and the presence of simple and compound spores. Barley and Hop washings are usually found to contain fragments and spores of this mould (see Chapter X.) The mere presence of the mould indicates doubtless, a poor class of Barley or Hop, as the case may be. In most Barleys these appearances are probably emphasized during the sweating in stack. With hops, the growth would be facilitated undoubtedly by imperfect curing and damp storage, especially the latter. Mould of any kind, in Barley, is generally more or less evident by smell, excepting when the samples are kiln-dried. As in the case of other moulds associated with Barley or Hops, there is probably little to be feared from the actual growth retaining any vitality throughout the Brewing process (for Copper-boiling must mean practical sterilization) ; but what is to be feared is the bad quality of materials showing mould, and the deterioration that must ensue in the structures of both barley and hops, from even a limited growth of such organisms, apart from the unpleasant flavour which they always impart to the material on which they subsist.

Mould spores of many kinds are so generally diffused in the atmosphere, that they are often found attached to the exterior of perfectly healthy vegetable products, and given the necessary conditions such as damp, etc., it is not long before growth renders their presence evident to the unassisted eye.

Besides the Black mould mentioned, there is no doubt that Hops are, in the "gardens," subjected to the destructive influence of other varieties : the so-called Hop Mildew, Sphærotheca Castagnei, is probably the commonest. This mould eventually forms black patches, and may be the same as that we have already described as an Ustilago.

We once had an inferior sample of hops in our hands that, whilst in bale, had developed a yellow mould curiously resembling the "condition" of the hop, the spherical sporangia being about the same size as the resin capsules. It is probable that this was the Eurotium form of Aspergillus Glaucus before referred to.

Where moulds develop on the surface of bales kept in a damp store, the infection is probably from the air, the mould growing on the damp sacking and spreading inwards to the hops. If the hops themselves are damp from undercuring or exposure, the growth would be doubtless facilitated.

FUSARIUM HORDEI, the red mould described by one of us some years ago,* is, after Penicillium Glaucum, the most frequently occurring mould in connection with growing Barley. It is occasionally seen amongst inferior samples of Barley, appearing as a crimson or pink tinted patch on defective corns, usually at the germinal end; fortunately, it does not spread to healthy corns, but it may be communicated to crushed ones. The most marked phase of its development is the crescent-shaped compound spore [Plate VIII., Fig. 5, a and d]. During its growth it may exhibit the following appearances :—Mycelial and aerial hyphæ, sometimes forming internal spores which escape from the end of the hyphæ, and may be called pseudo-spores [Fig. 5 b]; Fasces, or bundles of crescent-shaped spores on very short, thick hyphæ [Fig. 5 a]; Sporangia on lateral and terminal or long hyphæ [Fig. 5 c].

The presence of F. Hordei usually indicates a poor Barley. When submerged in beer-wort, the mould gives an alcoholic ferment form somewhat resembling Mucor Racemosus.

MONILIA CANDIDA is the name under which Hansen has described† a mould which, in certain phases of its develop-

---

* "Journal R. Micr. Soc.," Ser. II., Vol. III., p. 321.
† Carlsberg Report, Vol. II., part 4, 1886.

ment, shows cells resembling Saccharomyces, remarkable for the property they possess of causing, without previous inversion, alcoholic fermentation in a solution of Cane Sugar. The cells fall to the bottom of the liquid, multiplying like yeast, and may come to the surface again, forming a film.

The red growth of cells resembling Saccharomyces, observed by Hansen and others in beer-wort, is caused by organisms which probably bear a closer relation to moulds than to Alcoholic ferments proper.

If it be desired to cultivate any particular mould, various media offer themselves as favourable for the purpose :— Slices of boiled vegetables, *e.g.*, potatoes, turnips, etc. ; crushed germinating barley ; gelatine in small dishes : all or any of these may be sown with detached portions of the mould growth.

Spores may be grown experimentally in water alone, or on wet sand. The vessel containing spores or the mould growth, may be placed on a soup-plate containing a little water, and should be covered with a bell-jar or some suitable glass vessel, to exclude dust and secure a moist atmosphere. Many of the precautions described under Yeast and Bacteria may with advantage be adopted in pure cultivations of moulds.

CHAPTER VII.

The Bacteria or Schizomycetes.

I T is perhaps in connection with the Bacterial contamination of fermentable liquids that Pasteur's researches
have their highest value. In his "Etudes sur le Vin" many
of the disease changes to which the French red and white
wines are at times prone, are traced by him to their sources
in certain specific forms of Bacteria, giving rise to acidity
and unpleasant flavours. As a sequel to this he made
Beer his study, and by a succession of beautiful and
original researches demonstrated the fact, amongst others ·
that the changes involving perhaps the greatest loss to
which Brewers are subject, are those connected with the
growth of various kinds of Bacteria ; and that the exclusion
of these from the process by attention to various important
points, is one of the chief factors of success as regards the
product.

It has become the custom for some scientists of a more
modern school, to underrate the successful efforts Pasteur
made to place the whole Brewing process on a more stable
foundation ; but the fact should not be lost sight of, that
had it not been for his brilliant work there would still be
much groping in the dark in connection with the science
of Brewing ; for after eliminating from the process the

disturbing conditions due to Bacteria, he paved the way for a fresh departure as regards the study of the Alcoholic ferments.

We purpose in this chapter to give a general sketch of the Identification, Classification, Life-history, and Cultivation of Bacteria ; leading on to their connection with Brewing, and the effects due to their growth and action in fermented beverages, more especially in beer.

Some little confusion has arisen from the variety of names that has been applied to the Schizomycetes or fission-fungi, the words Germ, Microbe, Bacterium, Micro-organism, all indicating the same class of organisms. As in the case of the moulds there is no well-defined distinction as to form, between Alcoholic ferments (Saccharomycetes) and Bacteria, nor between Bacteria and Moulds ; the smaller size of the Bacteria in each case constituting the principal difference. It is hardly necessary to say that the microscope is the indispensable adjunct to all kinds of work on Bacteria, and the lenses of the instrument cannot be too good. For anything approaching to a *study* of Bacteria, magnifications of from 400 to 1,000 diam$^{rs.}$ are necessary, but for ordinary Brewery observations 300 diam$^{rs.}$ will suffice.

Some of the earliest observations of Bacteria were made about the year 1680 by Leuwenhoek, who in some letters to the Royal Society speaks of minute organisms in the lees of wine and beer, and also in putrid water, saliva, etc. Dr. Hooke had three years previously to this brought under the notice of the Royal Society, observations of his on small moving organisms in infusions of pepper and of other vegetable products. A certain Dr. King, working contemporaneously with Hooke, also observed and described minute organisms.

Seeing that 200 years have elapsed since these early investigators recognised and described what were doubtless

Bacteria, it is rather strange that not till the last 15 or 20 years should any very rapid advance have been made in this branch of scientific investigation. In recent times progress has been indeed rapid, owing to the great talent and skill brought to bear by men like Cohn, Koch, De Bary, Zopf, and Pasteur. The name of Cohn deserves more than casual mention, for he helped largely in laying the foundations of a scientific study of Bacteria, by very careful researches leading to an improved classification.

Before going into detail in connection with some of the members of this " Kingdom of the infinitely little," as it has been aptly termed, let us quote some every-day instances of effects produced by the development of Bacteria. We have the Souring of Wine, Beer, Milk, and other liquids ; the ripening of cheese, especially in the case of such powerfully-flavoured varieties as Limburg, Roquefort, Camembert, etc. ; the putrefactive decomposition of meat and fish ; and many other decompositions like those of " Brewers' grains " and " spent hops," where a bad smell is a noticeable feature. In all or any of the above-mentioned cases, Bacteria may be easily detected by the Microscope, and we recommend the student to obtain in the first place a general idea of these organisms. Slime from a dropping water-tap, or some steep-water from the maltings, kept for a day or two in a warm place, will usually furnish Bacteria in some variety as regards shape and size. The Bacteria appear in the form of round, or cylindrical rod-shaped (rarely fusiform or spindle-shaped) cells of very minute size. The diameter of round cells or transverse section of cylindrical ones is generally about $1\,\mu$; the length of cylindrical cells is not commonly more than 2 to 4 times their transverse section, although some cells may attain a diameter as great as $4\,\mu$, and occasionally grow to an enormous length.

8

We will now enter into consideration of the structure of Bacteria. The outer membrane is, as in the case of the Saccharomycetes and Thallophytes, a fairly resisting and elastic substance, probably of the nature of cellulose. It is free from Chlorophyll, and in the majority of cases colourless. The inner portion of the Bacterium cells is a pasty mass rich in Nitrogen, called Protoplasm or Mycoprotein, varying in density, transparency, and refractive power. Many Bacteria enter into a motile or actively-moving state at some period of their development; the organs by which this movement is effected being hair-like protrusions, known by the names of Cilia or Flagella, which having about the same refractive power as water, are only seen with difficulty, even when the movement ceases; but they can be rendered more distinct if the membrane of which they consist, be condensed by treatment with Iodine solution or Osmic acid. It is affirmed that some organisms have the power of retracting the Cilium into the cell. A curious point is mentioned by Zopf—viz., that Micro-photography will sometimes render visible the cilia that cannot be seen by the eye; the sensitized photographic plate being more susceptible than the retina.

We may now deal with the plan of reproduction or multiplication, of some typical Bacteria chosen from those associated with the Brewing process, and for convenience in grouping, take Cohn's classification of 1872.

Class I.—Sphæro-bacteria = Dot or sphere.
  „ II.—Micro-bacteria − Short rods.
  „ III.—Desmo-bacteria = Threads.
  „ IV.—Spiro-bacteria   = Spirals.

Class I., generally known as the Coccus or Micrococcus form, propagates usually by simple division in two or more directions; the segments thus formed enlarge as the fission progresses, and as they arrive at maturity are liable to become disassociated from each other. This is very well shown by an organism called Sarcina Litoralis, found in putrefying sea water and spring waters. The organism is shown on Plate IX., Fig. 3, in successive stages of reproduction.

Another coccus form, but not a true micrococcus, is brought about by the breaking up of longer or shorter rod-lengths into ovals and spheres, by constriction of the outer envelope at given points. Separated pairs of cocci formed in this way are called Diplococci. Bacterium Aceti, B. Pasteurianum, and B. Xylinum, afford at certain periods of development, excellent examples of diplococci ; as shown in Plate X., Fig. 4. Budding of one spherical cell out of another has not, so far as we know, been observed in connection with Bacteria.

The normal methods of reproduction of Classes II. and III. are—(*a*) By the continuous development of rod lengths ; (*b*) By the formation of spores capable of germination, and consequent reproduction of the Bacterium form. The organism Cladothrix dichotoma, affords a good illustration of rod lengths, showing at the same time what is called false-branching (see Plate IX., Figs. 1 and 2). As an example of spore formation Bacillus Subtilis may be referred to [Plate IX., Fig. 4].

The Spirillum forms of Class IV. are reproduced by fission, sometimes in short lengths, sometimes in very long pieces which afterwards break up into separate individuals.

According to Cohn, all Bacteria tend to reproduce a constant and uniform type ; Micrococcus yielding Micrococcus, and Spirillum, Spirillum : but in the light of Zopf's more recent researches this position is no longer tenable, for he shows clearly that in the case of many kinds of Bacteria, long rod lengths may produce short ones, and even coccus forms. In fact almost every form assumed by

Bacteria in general, may be furnished by one organism, as for example, with Cladothrix dichotoma [Plate IX., Fig. 2], which shows coccus, rod, and spiral forms. It was only to be expected that Zopf, after satisfying himself that distinct species of organisms could go through a cycle of changes—at some period in which they might have departed from the typical form, and were not to be morphologically identified—should have devised a classification to cover the differences of form he had encountered. The following is his somewhat elaborate system :—

Class I.—Coccaceæ—Micrococcus forms, and threads of cocci.

„   II.—Bacteriaceæ—Cocci, short rods (Bacteria), long rods (Bacilli), long threads (Leptothrix) ; no spirals.

„   III.—Leptotricheæ—Cocci, Bacteria, Bacilli, Leptothrix, and Spirals.

„   IV.—Cladotricheæ — Cocci, Bacteria, Leptothrix, Spirals, and false branching.

The first three classes include the forms mentioned by Cohn ; class I. being the same as his. Criticising the arrrangement, we feel inclined to remark that it was hardly worth while establishing a new class for the organisms exhibiting false-branching.

A very elaborate classification of the Schizomycetes has been devised by Flügge, but space will not allow us to introduce it in detail. It contains two general groupings into round and ovoid cells and cylindrical cells, with about twelve different subdivisional groups. Very slight morphological differences are made of much importance, and generally speaking, the classification seems a cumbrous one. Nägeli includes the whole of the classes in the one term Schizomycetes, and maintains that Bacteria are allied to yeast. He classes all the microscopic fungi producing decomposition as follows :—

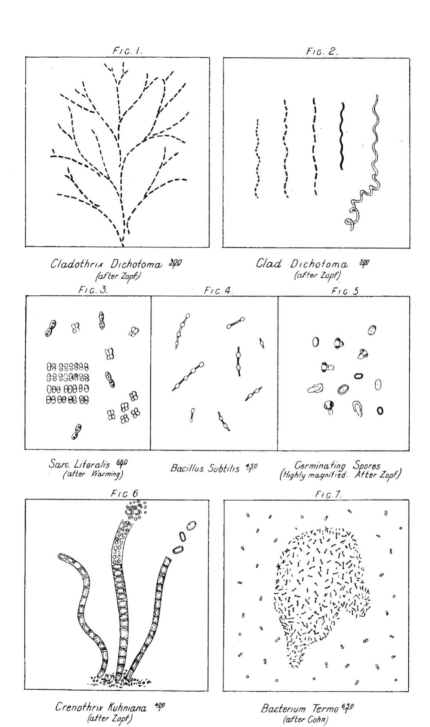

FIG. 1.

Cladothrix Dichotoma $\frac{300}{1}$
(after Zopf)

FIG. 2.

Clad Dichotoma $\frac{300}{1}$
(after Zopf)

FIG. 3.

Sarc. Litoralis $\frac{660}{1}$
(after Warming)

FIG. 4.

Bacillus Subtilis $\frac{450}{1}$

FIG 5

Germinating Spores
(Highly magnified. After Zopf)

FIG 6

Crenothrix Kuhniana $\frac{400}{1}$
(after Zopf)

FIG. 7.

Bacterium Termo $\frac{650}{1}$
(after Cohn)

Mucorini or Moulds.

Saccharomycetes or Alcoholic ferments.

Schizomycetes or Bacteria.

It then appears that in the majority of cases it is almost impossible to identify a Bacterium from its mere appearance at any given time. Zopf's statements to this effect are supported by Klebs and other workers who have seen rod and spirillum forms produced by the same organism

There seems to be a very general tendency for Bacteria in the form of rods and threads, to become curved or crooked, especially with alterations of nourishment. Besides the normal forms exhibited by Bacteria, very curious deformities are occasionally met with, showing dark coloured protoplasm and marked peculiarity of form, including great enlargement of certain cells ; insomuch that were the portions viewed alone, one would not associate them with the original Bacteria. Such deformities are called Retrograde or Involution forms, and are probably brought into existence by deficient nourishment.

Let us return to a closer consideration of the reproduction by fission, or division of rod lengths. Zopf says that the membrane by which the Bacteria are enveloped is in many cases capable of thickening, and then dividing into layers, one of which (the inner) is capable of differentiating itself, whilst the other (the outer layer) grows for a longer or shorter time, till finally it may yield to the pressure of enclosed cells and some of these last be pushed out, as for example with Crenothrix Kuhniana—shown in Plate IX., Fig. 6—an organism which is somewhat closely allied to the Mucorini, but classed by many authorities amongst Bacteria : it also affords a striking example of cell formation, by differentiation of the protoplasm in the threads.

We can now proceed to consider in some detail the

reproduction of Bacteria by spore-formation, a process first observed by Cohn in Hay Bacillus (Bacillus Subtilis) [Plate IX., Fig. 4]. Spores are formed by a condensation of cell protoplasm into small spherical masses with a new and independent membrane. A disintegration of the old cell envelope often takes place about the same time, thus freeing the spores. The process of sporulation is plainly seen in two kinds of Bacteria, the Bacillus Subtilis already referred to, and the Butyric ferment (Bacterium Butyricum) [Plate X., Fig. 7]. Sporulating Bacteria may usually be found in decomposing Brewer's grains, or in hay-infusions or steep-water kept at 80° to 90° F.,—in the latter cases the sporulation usually takes place after an active growth of the Bacillus ;—they are seldom met with in Beers, but we have once or twice seen what was apparently Bacterium lactis in the sporulating state. In Plate IX., Fig. 5, we give examples of germinating Bacterium-spores, very highly magnified.

Bacteria under certain circumstances, generally those of restricted growth, develop a very curious condition known as the Zoogloea or resting state, caused probably by the gradual reproduction of Bacteria in close proximity, and the tendency the organisms then have to largely increase the enveloping material, which at the same time passes into a gelatinous condition. This may proceed until the contour of the separate cells is nearly lost, and an almost indistinguishable mass may remain, where formerly well-defined Bacteria were seen. Sporulation may go on at the same time. Plate IX., Fig. 7, shows Bacterium Termo in the zoogloea form.

To proceed with some of the more general phenomena associated with the growth of Bacteria. The production of a definite pigment is a property belonging to a fairly large class called Colour-bacteria. Amongst the colours produced (which are usually diffused in the cultivating medium) are Crimson, Blue, Scarlet, and Yellow. The Bacterium form is in most cases a small sphere. Boiled white of egg is an excellent nourishing material for these growths. A great variety of products is obtained from Bacterial decomposition ; amongst the commonest are free acids such as Formic, Acetic, Lactic, Butyric, and other organic acids, formed from a variety of substances, viz., alcohols, glycerine, vegetable gums, starchy bodies and the Carbohydrates generally; or Ammonia may be produced from certain nitrogenous bodies, especially amides and albuminoids ; or an oxidation of the Ammonia to Nitric and Nitrous acids may ensue—a highly important action that is always going on in porous soils charged with sewage and decomposing animal and vegetable matters, thus bringing them into a form in which they can be assimilated by plants, and so enter again into the round of life. Hydrogen, Nitrogen, Sulphuretted Hydrogen, Marsh gas, Phosphoretted Hydrogen, and Carbonic acid gas are also products of decomposition by the intervention of Bacteria.

Nägeli has advanced the same theory in connection with Bacteria that he holds respecting Yeast, viz., that the decompositions are set up by molecular vibrations of the protoplasm, which are communicated to substances within a certain radius of the organism.

In the case of all Bacterial decompositions, a point is reached when the action of the organisms is arrested or checked by the nature of the products, which act towards them as poisons, whether they be acids, alcohols, or other of the substances mentioned. Doubtless a sudden immersion of Bacteria in solutions containing even less of those sub-stances than is normally produced, would arrest the activity of the Bacteria or kill them. The same thing holds to a certain extent in the case of the Moulds (more especially in a submerged state) and the Saccharomycetes.

Bacteria thrive best in weakly alkaline solutions, containing Carbon and Nitrogen, etc., in the form of Carbo-hydrates and Albuminoids. After the Bacteria have ceased growing, Saccharomyces forms and moulds may appear. In acid solutions, such as Wine-must and Beer-wort (which last has a slight normal acidity), the sequence is different, the Saccharomycetes developing preferably, followed by Bacteria, Aerobic ferments, and lastly moulds.

Bacteria have a considerable affinity for Oxygen gas, especially when they are in the motile state. This is shown in an interesting way when a cover glass is placed over a drop of liquid containing Bacteria, on a microscope slide, air bubbles being also enclosed : the moving Bacteria flock to the edges of the bubbles, and also to the edge of the cover-glass.

With regard to the action of electricity on Bacteria :—a weak current affects them but little, but a stronger one can sterilize a solution in a time proportionate to the strength of the current ; sterilization being more complete at the positive pole of the Battery. The killing of the organisms present, does not however, prevent newly sown Bacteria subsequently developing. It is probable that Spores would not be killed by even a powerful current of electricity.

Very little exact work is extant in connection with the effect of chemical substances on Bacteria. It appears that mineral and fruit acids, and some other organic acids, have a marked deterrent effect ; this is notably the case with Bacillus Subtilis, which is hindered in its growth by even a weakly acid solution. Sulphurous acid and Salicylic acid have a well-marked antiseptic effect on bacteria, and we shall have occasion to make some further remarks on their use.

Many substances having a markedly destructive action on developed Bacteria affect the spores of the same but little, for they have been dipped in concentrated solutions of Sulphate of Copper and Mercuric Chloride without losing their capacity for germination. It is also well known that the spores of some bacteria, *e.g.*, Bacillus Subtilis, can withstand exposure to a boiling temperature for a limited time without sacrificing their vitality.

Gradual and nearly complete loss of water may be sustained by bacteria without loss of vitality; desiccated spores of some bacteria retaining their power of germination for several years. Forms other than Micrococci or Spores do not however, live long in a dry state.

Temperatures below 60° F. are unfavourable to the development of bacteria, but between this temperature and 130° F. each kind of bacterium finds some very favourable range.

It will be desirable to consider in some detail the chief methods of research adopted for Bacteria. In connection with the growth and culture of these organisms, the following important questions present themselves :—

1.—In what plasma or food stuff does the Bacterium thrive best?
2.—Through what stages of development does it go?
3.—As to the products of decomposition by specific bacteria.
4.—The behaviour of the organism in relation to Oxygen.
5.—The influence of temperature.
6.—The action of antiseptics.

We shall in some way or another touch on nearly all these points. At present we purpose dealing with modes of cultivation :—

Having obtained a fair example of the particular kind of bacterium one is desirous of cultivating, the next thing is to provide a suitable plasma and keep out air-borne germs. Amongst the various cultivating media there is nothing much better for bacteria than gelatine, which can be adapted for use in a variety of ways; the addition of meat-peptone or meat extracts, such as Liebig's, Brandt's, etc., undoubtedly increase the nutrient power. (See Appendix C 1.)

Space will not permit us to enter into the minutiæ of the sterilization of vessels, infusions, etc. In all cases it is a matter of employing heat in such a manner as to kill germs that are not wanted.

For bacterial research, a room that is as far as possible free from dust is desirable. Glycerine when smeared on plates, and on the inside of Bell-jars used in connection with cultivations, will arrest floating dust in an effective manner. A supply of wide-mouthed bottles, test-tubes, and flasks for a stock of gelatine, with a few other small pieces of apparatus, will enable one to grow ordinary bacteria. Such things as sand-baths, water-baths, drying-ovens, incubators, etc., can with a little ingenuity, easily be contrived out of every-day appliances, if one does not care to incur the expense attending their purchase.

There are several methods of obtaining a pure cultivation of one particular organism. Speaking generally, they commence with the excessive dilution (described in Chap. V., page 63), of a liquid containing a preponderating quantity of the organism sought. The Ranvier or Böttcher moist chamber may be used with gelatine; or "plate cultivation" (which has of late years been developed in connection with water analysis) may be carried out with the same medium, as follows :—

A portion of the liquid containing the organisms it is desired to cultivate, is withdrawn by a sterilized pipette and run into sterilized gelatine-peptone (Appendix C 1), which has been melted in the test-tube containing it, by immersing in a water bath at 86° F. Complete mixture is effected by shaking, and the fluid is run on to a clean, sterilized, uncoated photographic plate, resting in a perfectly horizontal position (secured previously by use of a spirit level) on a glass tripod standing on a soup plate. The arrangement is immediately covered by a glass shade. A 2 % solution of mercuric chloride, standing to a slight depth in the soup-plate, acts as an antiseptic seal. The glass plates are preferably sterilized by heating in a shallow metal box ; the rest of the apparatus by rinsing with mercuric chloride solution.

The whole arrangement is next placed in a chamber maintained at 58°—77° F. for incubation, which extends over three to five days. The plates are daily inspected, without removing the glass cover, and the appearance and growth of any colonies watched ; before these last coalesce, the plates are withdrawn for microscopical examination. The points to be noted are—

(1) The number of colonies—conveniently ascertained with a hand-lens and a superimposed glass plate ruled in equal squares.

(2) The effect of the growth on the gelatine itself.

(3) The nature of the organisms forming the colony, as ascertained by the microscope.

Marked peculiarities are at times, met with in gelatine cultivations ; for instance, the formation of gas bubbles, usually of lenticular shape, but varying according to the density of the gelatine ; and the appearance of spherical or pear-shaped liquid cavities, whose contents are usually acid.

A greatly improved definition under the microscope of the membranes, and of the internal parts of Bacteria, is obtained by staining with various dyes, a process very elaborately described in some of the works on Bacteriology. Staining in a simple form may be carried out as follows :— Two or three drops of the liquid on a slide (free, if possible, from matters other than bacteria) are dried off gradually on a metal plate, at a temperature of about 90°—100° F. A drop of very weak Rosaniline or Methyl-violet solution is put on to each of the dried spots, and evaporated to dryness as before. The surplus dye can be removed with weak alcohol or dilute nitric acid, which are in turn washed away with a little water, and some clove oil can be put on the spots, followed by Canada Balsam (Appendix B).

If the slide be wanted for immediate examination, it is unnecessary to remove the surplus dye; in such a case a little turpentine, followed by clove oil, will give very good specimens. Eosine is a good dye for immediate examination, but fades on keeping. We have already alluded to the use of Iodine and Osmic acid for rendering bacterial structures more plainly visible: they are both more suitable as mere reagents than for permanent specimens. The so-called pathogenic organisms, or bacteria associated with diseases of men and animals, interesting though they may be, are of course outside our province.

It is a matter of certainty that Bacteria have existed on this earth from some very distant period of its history: Van Tieghem found them in the fossil roots of Coniferæ and other fossilized vegetable remains from the coal-measures.

The relationship of Bacteria to the Moulds would seem to be much stronger than to the Alcoholic ferments, turning on this point more especially, that the Saccharomycetes do not exhibit any motile forms provided with Cilia; whereas in certain stages of their growth the moulds do. All three classes of organisms may form spores, but the Saccharomycetes do so under somewhat exceptional conditions.

Bacteria, as a rule, require more complex forms of nourishment than Moulds or Saccharomycetes; for the two last-mentioned will thrive fairly well in a mineral solution in which the carbon and nitrogen are represented by ammonic tartrate, whilst Bacteria grow but feebly in the same. Bacteria then may be ranked higher than moulds or alcoholic ferments, for although they show characteristics which at each end of the scale join them to separate kingdoms, making it a question between animal and vegetable life, the weight of evidence seems to us to point out a much closer alliance to the former than the latter condition of existence. Before proceeding to discuss the various forms of Bacteria that are associated with fermented liquids, a few words as regards the immediate source of these organisms will not be out of place. It is to the air we have to look, and it is not difficult to account for their presence in it, as Bacteria are constantly being produced in myriads by all kinds of decomposition of animal and vege- table substances in free contact with air. The products of decomposition or putrefaction becoming dried up are, with the myriad organisms and spores that they include, spread broadcast, the infinitesimal weight and size of the bacteria and their spores causing them to be carried to great distances, and easily kept in suspension by currents of air. In all populous districts Bacteria are everywhere, and on pretty well everything, and as a consequence we swallow them probably by thousands daily. Where the air is per- fectly calm, as in small enclosed spaces, bacteria settle down completely. Some interesting results relative to organisms in the air were obtained by Miquel from observa- tions made at Montsouris. In a cubic metre of air he found

In the Autumn	...	142 organisms.
,, Winter	...	49 ,,
,, Spring	...	85 ,,
,, Summer	...	105 ,,

In a cubic metre of air in the Rue de Rivoli, Paris, he found at one time 5,500 organisms. The experiments of Tyndall, Miquel, and others, have shown that the air at high elevations—for instance in the Alps—is free from Bacteria. We may then consider that the air is the reservoir from which the foreign organisms appearing in beer and other fermented liquids, are derived; though naturally we have to reckon on their possible and probable multiplication in the afore-mentioned media. Hansen, in the course of experiments that we have already alluded to, found that the air of Carlsberg contained a large number of organisms capable of growing in Beer-wort, and included forms of Bacteria that will be spoken of later.

We purpose dealing categorically with the Bacteria that may be associated with the process of brewing, and will group them thus:—

COCCUS OR MICROCOCCUS (including chains of Micrococci).
Sarcina.
The Viscous ferments.
Bacterium Aceti.
     ,,   Pasteurianum.
     ,,   Xylinum.
MICRO-BACTERIA (short rods).
Bacterium Termo.
     ,,   Lactis.
Pasteur's lactic ferment.
Bacterium Butyricum (also called Clostridium Butyricum and B. Amylobacter).
DESMO-BACTERIA (including long and short threads).
Bacillus Subtilis.
     ,,   Ulna.
Leptothrix.
SPIRO-BACTERIA (Spirals).
Spirillum Tenue.
     ,,   Undula

THE SARCINA GROUP.

Some very interesting and suggestive work has lately been published by Paul Lindner,* on this group of organisms,

* Nachrichten über den Verein Versuchs=und Lehranstalt für Brauerei in Berlin. Die Sarcina-organismen, etc.

which systematizes the information that was extant
before the experiments commenced, and adds thereto much
fresh matter. We will make some brief extracts :—An
organism of the Sarcina type but not a true Sarcina, seems
to have been alluded to by Pasteur,* as causing when
present in beer, a peculiar rough acidity and characteristic
odour. Bersch † mentions a definite Sarcina disease of
beers, causing a cloudiness which passed off in a few days
leaving the beer clear but with a disagreeable smell.
Hansen found Sarcina in yeast water : P. Lindner has
seen it in pitching yeast itself : and several observers,
amongst them Brown and Heron, have found it in malt ·
extracts. S. Von Huth has sought to establish a connection
between the fact of Sarcina growing readily in horse-urine
and stable-manure, and its appearance in beers : tracing
the contamination through air, water, ice, vessels, etc. He
maintains that Sarcina does not grow in liquids that readily
acidify, and notices that Sarcina-beers after a time lose their
characteristic taste and smell, but become vinous ; and
infers that when the development of acid reaches a certain
point the growth is arrested. P. Lindner, using Hansen's
moist chamber and gelatine mode of cultivation, with
decoctions of chopped hay (which favour the growth of
Sarcina), or malt extract solutions, and infecting these with
material from various sources, managed to separate and
identify several kinds of Sarcina, viz. :—

> Pediococcus cerevisiæ.
> Pediococcus acidi lactici.
> Pediococcus albus.
> Sarcina candida.
> Sarcina aurantiaca.
> Sarcina flava.
> Sarcina maxima.

* " Studies on Fermentation," trans. Faulkner and Robb. p. 6. Plate I., Fig. 7.
† " Die Bierbrauerei," 1881, p. 214.

Pediococcus cerevisiæ seems to be the one investigated by S. Von Huth, whose observations as to its presence in stable-manure are corroborated by P. Lindner : the latter has also seen it in well water used for cleansing purposes in a Brewery. It occurs in German beers, notably in Berlin Weissbier in the viscous condition, when it is probably the cause of the viscosity.

Pediococcus acidi lactici—probably the same as an organism seen by Hansen in the form of many-celled cubical packets or groups—gives rise to a considerable quantity of lactic acid. The diameter of the single coccus = 0·6 to 1 μ. Both this organism and Ped. cerevisiæ are by no means uncommon in German beer, and Lindner says that the intentional souring of worts in German distilleries is often carried out by Pediococcus acidi lactici.

Pediococcus albus was found in two spring waters ; resembles the foregoing forms ; it can give rise to a white pellicle.

Sarcina candida ; a form observed by Reinke in a Brewery water-tank ; gives brilliant white growths. Diameter of coccus 1·5—1·7 μ.

Sarcina aurantiaca ; produces an orange coloured growth on gelatine ; found in Berlin Weissbier. Mentioned by Frankel.*

Sarcina flava (de Bary); a Sarcina producing a yellow pigment, found during some of the experiments on Ped. cerevisiæ, previously described by De Bary, and probably the same as Schrœter's Sarcina lutea. Diameter of coccus, ·8 μ to 2—2·5 μ.

Sarcina maxima [Plate X., Fig. 1] ; packet-forms met with in malt mashes. Diameter 3—4 μ.

P. Lindner summarizes his observations thus :—

1.—The Sarcina group is represented by numerous kinds associated with fermentation. It is almost impossible to

---

* Grundriss der Bacterienkunde. 1887, p. 166.

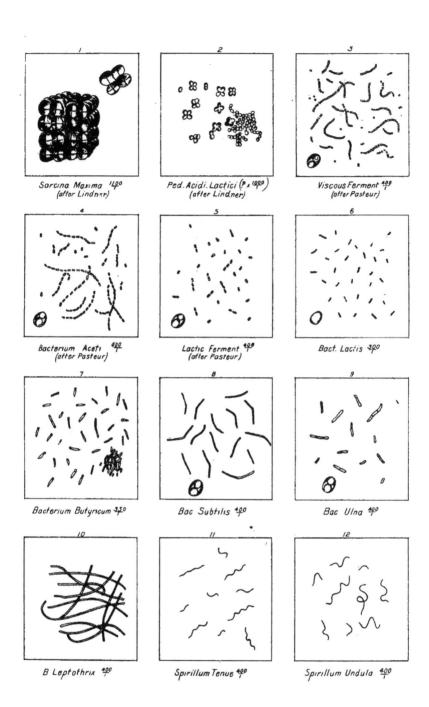

identify them by mere microscopical examination. Cultivation in different media is necessary.

2.—Some of the organisms show a two-dimensional growth, viz., Ped. cerevisiæ, Ped. albus, and Ped. acidi lactici.

3.—Others show a three-dimensional growth, but only in hay-decoction, they are Sarcina candida, S. aurantiaca, and a kind identified by Schrœter, S. rosea.

4.—Others grow almost exclusively in the typical Sarcina form—Sarcina flava and S. maxima.

5.—None of the varieties forms spores. P. cerevisiæ gives abnormal or involution forms. P. albus, P. cerevisiæ, and S. aurantiaca can form films.

6.—With the exception of S. maxima, which was not investigated in this respect, the different kinds produce varying quantities of lactic acid, with traces of formic acid

7.—Nearly all kinds liquefy gelatine sooner or later.

8.—A temperature of 60° C. (140° F.) kills any of them in a short time.

In addition to the kinds mentioned by Lindner, as investigated by himself and other workers, there remain some few forms of Sarcina which we will merely mention, as they have not, as far as we know, any traceable connection with the Brewing process. They are—

S. Reitenbachii (Caspary), found on water plants.

S. Hyalina (Kützing), in marshes.

S. Litoralis (already mentioned in connection with the mode of growth of Bacteria), found in spring water and putrefying sea-water.

Sarcina, as observed in English beers, is found in groups of four or tetracocci ; also as diplococci ; and may be disassociated into separate coccus forms. Sometimes the cocci are grouped symmetrically, at other times irregularly [Plate X., Fig. 2]. We have seen Sarcina not unfrequently in " forced " ales ; also in ales returned to the Brewer ; and lately we have seen some good examples of it in cask ales, some of which had become acid and vinous in store. The chief results of a free growth seem to be a high acidity—probably from lactic acid—sometimes preceded and accompanied by a vinous flavour, a harsh bitter, or else a peculiar woody taste. The Sarcina growth is generally accompanied by other bacteria. We are inclined to believe that two forms of Sarcina are met with in English beers :— One, a more symmetrical and less easily growing organism, probably Lindner's Pediococcus cerevisiæ; the other, in less symmetrical forms, appearing in greater quantity, and accompanied by acid production is, we think, Lindner's Pediococcus acidi lactici. Plate X., Fig. 2, furnishes, in our opinion, an example of the latter. We have made experiments to determine whence infection from Sarcina may proceed, and have convinced ourselves that very old wooden vessels constitute one source ; the organisms being not unfrequently discoverable in the spongy deteriorated wood. How they effected a lodgment there, and whence derived, are questions not so easily answered ; impure air or cleansing water, may in some cases furnish a solution. Apart from direct infection of beers which show Sarcina, there must be a predisposition to nourish the particular organism ; and as regards this point, the following remarks apply :—

We have encountered Sarcina in ales which were brewed with a large percentage of inferior moist sugar containing nitrogenous organic matter and phosphates. Various experimenters have shown that neutrality or alkalinity of nutrient solutions favours the growth of Sarcina ; a deduction from this being that reduction of the normal acidity of beer might be a predisposing condition. Amongst possible causes are included under-cured malt, especially if slack and otherwise of inferior quality. It is rarely the case that Sarcina gains any headway in English beers, though we have encountered it more frequently this year (1889) than at any previous time in the last twelve years ; and it would appear to have some direct connection with the character of the season's malt. Lager beers seem much more liable to Sarcina, owing very possibly, to the low temperature of malt-curing and the light hopping.

Viscous Fermentation,

Or the passing of fermented liquids into a viscous or "ropy" condition, is by no means an uncommon phenomenon. Peligot* seems to have been the first to notice a special ferment capable of producing it.

Pasteur† subsequently speaks of the viscous state in connection with wort and beer, and describes a special ferment [Plate X., Fig. 3], which is capable of transforming certain sugars into a kind of gum, together with Mannite and carbonic acid gas. Any acid formed such as Lactic and Butyric, resulting probably from other organisms present at the same time.

Viscous Beers are fortunately, comparatively rare. In most of the cases we have encountered, the quantity of the organism present seemed to bear a very slight relation to the effect produced. In one or two cases we have seen the organism in the chain form, but more generally in the coccus condition or with a tendency to form tetracocci, the latter fact rendering it probable that there is a relationship to one of the Sarcina forms described by Lindner, possibly Pediococcus cerevisiæ. The diameter of the cocci is 1.2 to 1.4 μ. By infection we have frequently excited marked viscosity in cane sugar solutions with very little clouding and with the production of an exceedingly small quantity of the organism, which strengthens our view that the

* Traité de Chimie de Dumas, vol. vi., p. 335, 1843.
† "Studies on Fermentation," trans. Faulkner and Robb, p. 5.

viscosity is more especially the result of some unorganized ferment eliminated by the Bacterium.

Very little of a definite character can be advanced as to the causes which favour viscous fermentation in Beer; it is probable that inferior and very slack malt, light hopping, and imperfect cleansing—owing to the nature of the worts and the weakness of the yeast—all tend to do so. Direct infection from cask seems to us quite a possibility, where ropiness only declares itself occasionally and not in connection with a whole brewing.

Various artificial solutions can be made, which favour the growth of Sarcina, for instance :—Yeast-water, made by boiling up yeast with water and filtering; aqueous extracts of wheat flour, barley and rice, with some added sugar ; and the liquids mentioned by Lindner, viz. : sweet wort, and a decoction made by treating chopped hay with boiling water. Neutralisation of any free acid in the solutions seems to materially aid the growth, especially if any acid subsequently produced be neutralized as it is formed, by introducing powdered chalk, marble, etc. According to Pasteur the amount of gum produced does not stand in constant relation to the sugar decomposed, and he therefore thinks that there are different viscous ferments, one of which forms only gum. At the present time there is considerable scope for investigation of viscous fermentation as there is comparatively little known about it. Before leaving this subject an interesting fact may be mentioned, viz. : that the phenomena of viscous fermentation are exhibited by an organism called Leuconostoc mesenteriodes, which has the power of converting large quantities of the juice of the sugar-beet into a mucilaginous mass, in a comparatively short space of time ; causing complete loss of the material. We merely mention this without wishing it to be inferred that there is any connection proper between Leuconostoc and the process of Brewing.

MYCODERMA ACETI,

Or Bacterium Aceti, as it is perhaps preferably termed, also popularly known as "Mother" of Vinegar, is the organism commonly associated with a change that alcoholic liquids are liable to undergo, during which the alcohol is converted into acetic acid, and this last subsequently into water and carbonic acid gas. The appearance of a film or pellicle on the surface of the liquid is a very ordinary accompaniment of its growth. Pasteur was the first to establish the known relation of the organism to its products : he showed, moreover, that if the action of the ferment was weakened, Aldehyde may be first produced from the alcohol ; the consequence of which would, as regards beer, be a vinous flavour. Acetic ether may also be produced at the same time and considerably enhance this effect.

On referring to Plate X., Fig. 4, the organism is seen in the characteristic chain and diplococcus form, the smaller dimension of the latter being about 1 μ. Bact. Aceti is coloured yellow by Iodine.

Bact. Aceti has been made the subject of a very careful investigation by Adrian Brown[*] who took all precautions to secure pure cultivations. He describes it as forming a greasy pellicle, inclined in the early stages of its growth to climb up the moist surface of the containing vessel. The liquid below the pellicle is usually turbid from suspended cells. In liquids free from oxygen it does not increase but keeps alive for a long time. It forms figure-of-8 cells 2 μ long, united into chains of varying length and sometimes the chains are composed of distinct cocci. Adrian Brown also observed abnormal or involution forms 10—15 μ long, and of a dark grey colour. The shorter rods and cells of B. aceti, when floating freely in the liquid, are motile.

[*] J. Chem. Soc. Transactions, 1886, p. 172, and 1887, p. 638.

Acetic acid is the one and only acid formed by a pure growth' of B. aceti, and it may be further decomposed into Carbonic anhydride and water, thus substantiating Pasteur's statements. Where alcohols other than ordinary or Ethylic alcohol, are present, B. aceti produces acids corresponding to them, and it seems to us that this fact is calculated to throw some light on the variety of flavours produced by the ageing and incipient decomposition of alcoholic liquids, which last may be considered as being more or less prone to the incursion of Bact. Aceti on exposure to the air, and especially so where the liquids are directly infected, as for instance, by an acid cask. The organism is made use of in vinegar factories; the liquids to be acetified being passed through vessels containing porous material, such as shavings, etc., strongly infected with the Bacterium. Liquids that contain over 10 % of alcohol do not allow this organism to thrive. Temperatures approaching 80°—90° F. are very favourable to its growth.

Bact. Aceti figures very often in ales returned in partially filled casks, the free exposure to air being the determining factor of its growth. In imperfectly-corked bottled ales it sometimes appears as a film, as also in defectively stoppered forcing flasks. A very moderate infection of B. Aceti will cause marked acidity and deterioration of ale; and there is no mistaking the presence of its product, acetic acid, with its highly characteristic flavour. With a normal process, and due attention to cleanliness of vessels —especially cask plant—there is comparatively little risk from B. Aceti.

BACTERIUM PASTEURIANUM.

Hansen* has given the above name to a form of Micro-coccus which has the same appearance as B. Aceti, and

* Meddelelser fra Carlsberg Laboratoriet, Andet Hefte, 1879, pp. 73 and 96.

like it, produces acetic acid: it is in fact, only distinguishable by its giving a blue colouration with Iodine, this characteristic displaying itself however, through successive generations.

BACTERIUM XYLINUM

Is an acetic ferment which forms cellulose. It was discovered by Adrian Brown, and described by him* as being identical with the so-called vinegar plant. It was grown in red wine diluted with half its bulk of water, and rendered acid with 1 % of acetic acid in the form of vinegar. Beyond the production of acetic acid, the main peculiarity in connection with the growth of the organism is the formation of a surface membrane of cellulose, which if shaken down, is renewed time after time, and appears to be the only form in which the ferment develops, though the membrane may in some cases be dispersed through the liquid, giving a jelly-like appearance.

Microscopically, the organism exhibits itself in lines, embedded in a transparent, structureless film. The bacteria are most commonly rods about 2 μ in length, several often being united together. It is sometimes seen in a micrococcus form, which Adrian Brown suggests may be spores; also in long twisted threads, 10—30 μ in length, of a Leptothrix nature. It does not exhibit the large swollen involution forms of B. aceti. A temperature of 28° C. (82·4 F.) is most favourable for its growth. Gives rise to the same chemical changes as B. aceti. The formation of the membrane constitutes the chief difference between the two organisms.

PASTEUR'S LACTIC FERMENT,

Shown in Plate X., Fig. 5; occurs as a small rod bacterium generally contracted in the middle, giving somewhat of a figure-of-8 shape. It often occurs in short

* J. Chem. Soc., 1886, Trans., p. 432.

chains of 2 or 3 individuals.   It is a question whether this bacterium is the same as the short rod form seen in beers, which we are accustomed to regard as B. lactis, but for our present purposes it will be convenient to consider them under the same title.

By lactic fermentation is understood the transformation of certain substances into lactic acid, the presence of which in liquids becomes evident by a sharp acidity not necessarily accompanied by any distinct flavour, as in the case of acetic acid.   When milk turns sour spontaneously, the sugar it contains is converted into lactic acid, and it was from this source that the acid was first extracted.   It would appear that the presence of nitrogenous albuminoid matter is, in addition to sugar, required for lactic acid fermentation.   The temperature most favourable to action is 120° F., and the souring of a liquid such as wheaten flour and water, goes on with extraordinary rapidity at this temperature, a large amount of acid being formed before the action is arrested. By neutralisation of the liquid with chalk, etc., a much larger quantity of the acid is produced.   B. lactis is said to be able to grow without free oxygen : if it does so, it is probably only to a limited extent in comparison to its growth with free access of oxygen.

The German distillers believe that a small percentage of lactic acid in the worts secures a more vigorous fermentation, and one less likely to develop bacteria.   The presence of lactic acid is secured by exposing a small green-malt or other mash, infected with lactic ferment, to a process of souring for many hours at the favourable temperature 120° F. ; it is then mixed with the mash proper. It is more than probable that a considerable variety of organisms produce lactic acid ; thus it will be remembered that Lindner has observed lactic acid production with organisms of the Sarcina group, especially with Ped. acidi lactici.

It is generally assumed that a small quantity of lactic and acetic acids is always present in beer; we do not think, however, that the free acid of beer necessarily consists of these. Bact. lactis as seen in beers is generally in the form of small rods, 2 to 3 μ in length (see Plate X., Fig. 6), and sometimes in threads containing from 2 to 5 individuals; it is not certain, however, that this form is B. lactis. The single rods are often motile.

Bacterium lactis is the most commonly occurring disease-organism encountered in the brewing process, for it is exceptional to meet with beers and yeasts that do not show an individual here and there when submitted to microscopical investigation; and in most breweries it is discernible at all times in varying quantity. The degree of risk attending its presence depends mainly on the destination of the ales; that is to say, whether they are for "stock" or for a "quick" trade; for objectionable as bacterial contamination in a brewery is, there is a much greater margin for it in the latter case than in the former, the beers not having time to turn sour unless the contamination and yeast deterioration are so marked as to place the source of the existing trouble beyond a doubt. In the case of stock ales it goes without saying that too much care cannot be taken to ensure freedom of yeast and beer from B. lactis or other disease organisms. With the present method of brewing there *must* be contamination in various ways, but it may be reduced to a minimum by careful selection of yeast, and by **due** attention to the process; and with proper precautions, beers of such character can **be** brewed, that the few Bacteria remaining in them are almost inert, normal secondary fermentation being the only change.

BACTERIUM TERMO.

A small cylindrical bacterium about 1·5 to 2 μ long,

having a central constriction, giving it somewhat of the diplococcus or figure-of-8 appearance. It is about the commonest accompaniment of rapid putrefaction and decomposition, especially in meat infusions. It is actively motile, having a cilium or flagellum at each end. This organism was investigated by Cohn, who describes amongst other things the well-marked Zooglœa state, which it enters into [Plate IX., Fig. 7], alluded to earlier in this chapter. Bact. termo can multiply with enormous rapidity. In its ordinary state it is seldom noted in Beer and yeast, but may be found in the slime of pipes, accumulations in the corners of fermenting and cleansing vessels, etc., etc. It is possible that when present in worts its habit and form become somewhat modified, rendering it perhaps similar in appearance to B. lactis.

BACTERIUM BUTYRICUM,

Known also as Bacillus amylobacter and Clostridium butyricum, is an organism consisting of short cylindrical or slightly ovalled rods of somewhat varying length, their smaller dimension being about 1 μ. [Plate X., Fig. 7.] Motile forms have been observed by Pasteur, who observed also that the organism sometimes formed chains composed of the smaller individuals : he also investigated the chemical functions of the organism. B. butyricum very readily enters into the sporulating state, forming well-defined highly refractive spores, which as the original cell wall of the bacterium shrinks in and disappears, show up very plainly. At certain stages of its growth the organism may give a blue colouration with Iodine.

As its name implies, this bacterium is commonly associated with fermentations in which the production of butyric acid is the main feature ; the presence of the acid being declared by the peculiarly disgusting odour which is one of its attributes. Butyric acid is produced from substances capable of undergoing lactic fermentation, *e.g.*, Sugars, Carbohydrates, Fruit-acids, and Albuminoid substances. The temperature favouring its action most, is about 100° F., but like other organisms, it will grow at temperatures somewhat above, and considerably below this point. The organism must be pretty liberally dispersed in the air, as solutions of Cane Sugar, with the addition. of a little phosphate of soda or potash, usually develop Butyric acid when placed on the forcing tray (see Heisch's test, Chapter X.)

As regards the connection between Bacterium butyricum and the process of Brewing :—The organism is not discoverable in the yeast and beer associated with a normal process. It may however, be present in greatly deteriorated yeast, but is difficult to identify. Its presence in some stinking returned ales is indubitable, but this can hardly ever arise from circumstances that the Brewer is able to control; that is to say, it is not usually connected with any fault in the process, or if so, the fault or faults must be glaring indeed. Other examples of this organism are furnished by putrid grains and decomposing spent hops. The growth of B. butyricum appears to be arrested by a very moderate development of Butyric acid, but the extraordinarily powerful and disgusting smell of the latter renders *traces* of it plainly evident. If, as in the case of B. lactis, the acid produced is neutralized by chalk or marble, as formed, it gives rise to a large quantity of a corresponding salt, calcium lactate or butyrate as the case may be. Small quantities of either Lactic or Butyric acid—1·5 % of the former and ·05 % of the latter—retard alcoholic fermentation.* The figures if correct show, that butyric acid exercises a far more powerful effect than lactic acid. Acetic acid occupies an intermediate position in this respect, as ·5 % retards alcoholic fermentation.* It is

* Märcker : " Spiritusfabrikation," p. 493, *et seq.*

probable that organisms other than B. butyricum may give rise to butyric acid.

BACILLUS SUBTILIS.

Synonymous with Ehrenberg's Vibrio Subtilis and Cohn's hay-bacterium. It is usually seen as long rods, straight or somewhat curved, the width of which is about 1 μ and the length from 6 μ upwards [Plate X., Fig. 8.] In a free growth the rods exhibit wavy and other movements, being provided with a flagellum at each end. They enter readily into the sporulating condition, as mentioned earlier in the chapter.

A pure growth of Bac. Subtilis may be obtained by raising an aqueous decoction of hay to boiling, plugging the flask with cotton wool, and putting aside in a warm place ; the spores of Bac. Subtilis survive the treatment.

There can be little doubt that this organism is found in association with beer and yeast, as the result of an improper process. It is to be seen in racking beer sediments, barms, forced ales and returned sour ales, and the motile form may be sometimes observed. We have seen one or two doubtful cases of sporulation in forced ales. According to Cohn the organism produces butyric acid, but this has been disputed by other observers : it seems to us probable that lactic acid is one of its products and possibly butyric acid under exceptional conditions; but there is no active production of either in beer. The presence of the organism in beer is no doubt connected with the following conditions :—

a. Direct aerial contamination, especially in the autumn, when the air is teeming with germs.
b. Uncleanliness of plant and process generally.
c. Deteriorated store yeast and a faulty process, including wrong temperatures, etc.

It not uncommonly appears in quantity in some

breweries during the summer and autumn, and must be met with extra care, and attention to salient points like those above-mentioned.

Forced samples of ale sometimes exhibit " fields " swarming with this organism. In these cases it is rather curious to observe that there is not always a degree of acidity corresponding with the growth, the deficiency of oxygen may have a connection with this ; its appearance would, nevertheless, cause one to be very suspicious of the stability of the beer.

BACILLUS ULNA,

Discovered by Cohn, occurs in long or short, but very broad-cylinders or threads, 2 μ broad, and in a free growth as much as 10 μ long. [Plate X., Fig. 9.] It is found in certain infusions, such as of white-of-egg. We have obtained it as a fortuitous growth in gelatine cultivations, in which it formed liquid cavities. It is occasionally to be found in beers and yeast, in which case we usually ascribe it, for ascertained reasons, to dirty vessels and pipes. It does not appear to grow in beer, at least to any extent, and may, we think, be regarded simply as an index of uncleanliness. In some cultivations it seems to differ comparatively little in form from B. Subtilis ; it should however, we believe, be regarded as an essentially different organism.

BACILLUS LEPTOTHRIX

Occurs in long threads, which are sometimes of great length, and twisted on themselves. [Plate X., Fig. 10.] It is found in liquids such as putrefying sweet wort, etc., and in decomposing masses such as the slime that collects in wort- and water-pipes, etc. We have seen it in racking beers, into which it probably found its way from dirty vessels. It is possible that it is only a particular form of Bac. subtilis.

Spirillum Tenue and Spirillum Undula.

The spirillum forms, though common in rapidly putrefying liquids and moist masses, are however, in our experience, uncommon in connection with Brewing. We have seen Sp. tenue in returned sour ales, and once or twice in forced samples; and both forms in spontaneously decomposing sweet wort, and in waters treated by Heisch's test. Spirillum undula we have also seen in putrid grains, and slime from pipes and dripping water-taps. The morphological differences between Sp. tenue and Sp. undula are so very slight, that many observers regard them as the same species. Plate X., Fig. 11, represents Sp. tenue, which is about 1 μ thick, and 4—15 μ long. The same plate, Fig. 12, shows Sp. undula, about 1.4 μ thick, and 8—12 μ long; it has wider spirals than Sp. tenue, and an active movement, at times, by means of flagella.

A few other kinds of bacteria have been observed by Hansen* as appearing in malt worts exposed to air infection. They are—

Bacillus ruber (Frank.)

Bacterium pyriforme.

,,   fusiforme (Warming).

,,   Kochii.

,,   Carlsbergense (resembling B. butyricum).

They do not appear to us, however, to call for more than passing notice.

We will now touch briefly on the subject of Antiseptics from the general point of view. The substances most noxious to bacteria seem to be Chlorine, Bromine, and Mercuric Chloride, especially the latter; they are of course quite inapplicable to Brewing. Amongst the less powerful but still effective antiseptics are Salicylic acid and Sulphurous acid, with their various combinations. Sulphurous acid combined as bisulphite of lime is, as is

---

* Meddelser fra Carlsberg Laboratoriet. Andet Hefte, 1878, p. 73.

well known, of high value for cleansing purposes in the brewery, and also, but to a less extent, in the maltings. Other bodies exercising a well-marked antiseptic action are alcohol, common salt, alum, various metallic salts, tannin, creosote, carbolic acid, lime water, and thymol. One or two of these are naturally associated with beers, the remainder, however, are not so connected, and would in the majority of cases be quite unsuitable for cleansing plant: we mention them as having a specific effect on bacteria generally. Moulds, generally speaking, resist the action of antiseptics more than bacteria, and bacteria have greater resisting powers than the saccharomycetes.

We will bring this chapter to a close with a few hints as to the examination of beers and yeasts for bacteria. Average samples should in all cases be obtained, and many "fields" should be examined, the slide being moved systematically so as to constantly present fresh parts to view. The results of observation should be noted down, so as to specify in some way the number and kind of bacteria present; actual counting is sometimes out of the question. Note book terms may be applied to beer and yeast as follows:—
Clean. Moderately clean. Not very clean. Not clean. Whilst the quantity of bacteria may be represented arbitrarily by the numerals 1, 2, 3; anything over the standard of 3 being marked, *Quantity.* In the note-book, positions may be allotted to different kinds of bacteria. The following entry serves as an example :—

Beer $\frac{365}{11}$ Not very clean.   1 — 0 — 1$^{\text{ul.}}$

which we should interpret :—One of Bacterium lactis per two or three "fields." No Bac. subtilis, and 1 Bacillus ulna: the verbal description of course speaks for itself. As brewing processes usually vary so much in their state of cleanliness as regards bacteria, such a means of record as we have tried to describe must be adjusted or made relative to each process, the results not being exactly comparable.

CHAPTER VIII.

THE FORCING PROCESS.

W E have already in Chapters III. and IV. made frequent reference to Pasteur's classical researches into the fermentation of Beer and Wine, and we now wish to explain how the methods first employed in those researches, may with advantage be practically applied by the scientific brewer to the regular examination of his product.

Briefly, Pasteur's method of investigation may be said to consist of experimental fermentations with fermentable liquids which had been completely sterilized by repeated boiling in glass vessels, whose outlets were either shut off from the air, or so plugged with cotton wool as only to admit thoroughly filtered air. When the liquid in the glass vessels was found to remain free from change, it was inoculated with minute portions of the purest growths obtainable by the methods employed. We thus purposely define Pasteur's pure growths, because Hansen's recent work has shown conclusively that the separation of the Saccharomyces by shape alone is impossible, and it is therefore more than probable that many of Pasteur's experiments were conducted with more than one variety of yeast.

Forcing Tray in working order
(from a photograph)

One of the chief results of Pasteur's work was, that working under the conditions above-named, his fermented liquids usually remained free from Bacteria, and therefore free from those acid changes which under less favourable conditions are found to accompany, or more correctly speaking follow, alcoholic fermentation. Pasteur had very carefully investigated some of these Bacteria, more especially those producing Acetic and Lactic acids, and it was the consideration of his published works on Vinegar and Wine* that induced the leading Burton chemists to apply his methods of investigation to the systematic examination of Beer, prior to the publication of the well-known work " Etudes sur la Bière," in 1876.† One of the most earnest and indefatigable workers in this direction was Horace T. Brown, and he may be said to have first systematized the method of beer examination by " forcing."

It is to the consideration of this method of testing the keeping qualities of ales and worts—depending as it does so materially on the use of the microscope—-that we intend to devote this chapter; and in the first place we will describe the piece of apparatus employed first in Burton, and now generally used throughout England, known as the Forcing Tray [Plate XI.] It is an oblong vessel, made preferably of copper; the size varies somewhat, but we find the following dimensions very convenient :—2 ft. 9 in. by 1 ft. 9 in., and 3 in. deep; the upper surface may be turned up about $\frac{1}{8}$ in. all round to form a rim.

The want of attention to certain details, and to the fitting of various accessories, may very considerably vitiate any results obtained by the use of this apparatus ; so that we shall now describe in some detail the apparatus itself and the method of using it, before entering upon a description

---

* Etudes sur le Vin, 1866, and Etudes sur le Vinaigre, 1868.

† Translated by Faulkner and Robb, 1879.

of the varied microscopical observations, which are the principal sources of information gained by its use.

It is advisable to have an oblong sheet of thin copper or block tin, brazed or soldered on the under side of the tray, in order that the gases given off by the burner used for heating it, may not corrode the actual surface of the tray. When this protecting piece is found to be seriously corroded it may be easily replaced by a fresh one. Inside the tray it is usual to have another sheet of copper covering the central part, and supported about $1\frac{1}{2}$ inches from the bottom; this is called the Disperser, and extends to within about 3 or 4 inches of the sides of the tray; its purpose is to prevent the water directly heated by the burners rising at once to the top of the tray, and so causing an unequal heating of its upper surface. A fair sized tubular opening should be provided for filling the tray, and this may be loosely covered by a cap, or plugged with cotton wool. A tap or bib-cock may conveniently be fixed at one end or underneath for emptying when repairs are required.

The heating, which is all important, should whenever possible be with gas, as the regulation of any other source of heat is difficult. A piece of ordinary $\frac{5}{8}$-inch gas-piping about one foot long, with six porcelain-tipped nipples (size No. 2), screwed in at equal distances, is as suitable a burner as any (see illustration), and should the tray be square the gas piping may with advantage be bent into a circle. This burner is fixed so that the surfaces of the nipples are not less than three inches from the plate on the underside of the tray, and it is directly connected with the Regulator by india-rubber or "composition" tubing.

The usual form of Regulator is that known as Page's (Fig. 24), which may be described as follows :—

The bulb B and about an inch of the tube A is filled with clean mercury, which may conveniently be done by pouring small quantities of mercury at a time, into a small cone of stout writing paper with a good sized pinhole in the point, placed in the upper end of the tube A, or in the side tube K ; closing the opening not used, with the finger. The regulator is now placed in a small flask of water on the tray, as shown in Plate XI.

The tray being quite filled with water at a temperature sufficient to keep the thermometer on the tray about two

Fig. 24.　　　　　　　Fig. 25.

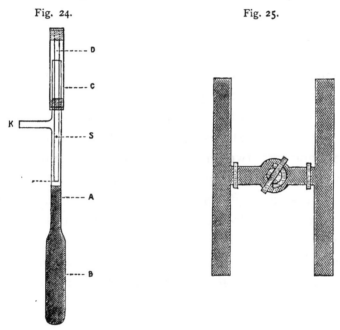

PAGE'S REGULATOR.　　　　METAL GAS CONNECTION.

degrees below the required temperature, a pint or two of water may be drawn off to allow for expansion. The sliding tube C of the regulator is now connected by india-rubber tubing to the gas main, and the outlet tube K similarly connected to the burner ; gas passes to the latter down the quill tube D, a small amount, sufficient only to keep the burner from going out, going direct through the pin hole S, the main portion through the end T— which in some cases is bevelled—and up the tube A into K. Under these conditions too much gas passes to the burners, and the temperature of the tray rises. When the thermometer on the tray indicates the required temperature (about 80° F.), the sliding tube C is pressed down until the lower end of the quill tube T just touches the surface of the mercury. If too much gas is thus cut off, the temperature of the tray falls slightly, the mercury in the regulator contracts and falls in the tube A, thus allowing more gas to pass through the quill tube at T, and so to the burners.

In an older form of this regulator, which is somewhat more reliable with a large amount of variation in the gas pressure (frequently the case in large works like breweries), the small hole in the side of the quill tube is replaced by a metal H piece (Fig. 25) with a tap in the centre; the two lower ends are connected respectively with the main and the burner, and the two upper ends with the quill tube and the side tube of the regulator. In use, the tap in the H piece is opened sufficiently to allow the same amount of gas to pass direct to the burner, as does the small hole in the previously described instrument; the remainder has first to pass through the quill tube of the regulator, the only passage being through the end T, which is directly controlled by the expansion and contraction of the mercury in the bulb. A modification of this arrangement is shown in Plate XI., the main gaspipe forming practically one limb of the H piece.

The flasks hold preferably about 120 cubic centimetres, and we find the pattern given in Fig. 26 more satisfactory to work with than any other; the old form with loose side tubes to be joined by india-rubber tubing being the cause of much waste of time, and therefore not to be recommended. In cleaning these flasks the greatest care must be taken to put no pressure on the side tube, as it is very liable to break off: well made flasks, however, stand ordinary handling well, and we have many that have been in regular use several years.

It will be as well now to consider the precautions necessary when collecting samples :—

Beer samples are best taken from the racking vessel · when beer is racked direct from cleansing casks, the samples may be taken from them or from the trade casks as soon as filled.

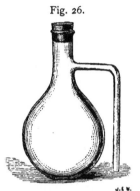

Fig. 26.

FORCING FLASK.

Ordinary 10 oz. stoppered bottles are quite suitable for collecting these samples, and may also be used for obtaining the sediment of the racking sample to be microscopically examined.

Each bottle should be carefully cleaned, and first rinsed out with the beer to be sampled, before filling ; it should also be labelled as soon as taken.

With regard to cleaning :—The bottles should be thoroughly rinsed round with a solution of caustic soda, then well washed out and allowed to drain neck downwards ; forcing-flasks may have a few cubic centimetres of a dilute solution of caustic soda boiled in them, followed by a thorough washing in clean water; draining as in the case of the bottles, taking place neck downwards.

A good draining-rack for forcing-flasks is made by stretching two stout copper wires about $1\frac{1}{2}$ to 2 inches apart over the sink. It is well to fill the forcing-flasks within twenty-four hours of taking the sample, and before filling they should be washed out twice with the beer to be sampled. When filled, the neck of the bottle is closed by a small India-rubber stopper which has first had some of the beer poured over it. It is better to leave about $\frac{1}{8}$ inch or so between the level of the beer in the neck of the flask, and the side outlet.

When placed on the tray, the side tube should dip not less than $\frac{1}{2}$ inch into mercury conveniently placed in small beakers, each of which will take five or six flasks standing round it [Plate XI.] Another method is to have the side tubes dipping into little troughs of wood or porcelain, filled with mercury; this allows of double rows of flasks on the tray, and utilizes a larger proportion of its surface.

The thermometer on the tray may be placed in a flask of water with some mercury at the bottom, so that the temperature indicated is practically that to which the beer in the forcing flasks is subjected.

As the object of these experiments is to see how far a beer may be expected to withstand the variations of temperature to which trade casks are subjected, the temperature at which these samples are maintained is in excess of that usually met with, and the growth of Bacteria and forms other than healthy Saccharomyces is thus considerably fostered.

The thermometer on the tray showing a constant temperature of 80° F., and the samples duly placed, certain observations may be made during the first few days; thus it is useful to note how long the beer takes to drop bright; if a large quantity of gas is given off quickly, or if the beer remains a long time before secondary fermentation commences.

Next, as to the length of time it is desirable to submit beers to this process. For Stock ales—and it is chiefly for this class of beers that the method of examination is of value—we find three weeks the shortest time practically useful, and where possible should advise four weeks.

When the flask is taken off the tray, the appearance of the sample should be noted, whether bright or not, and especially if there be any growth on the surface of the liquid. The liquid when cool should be decanted, and the specific gravity taken with a small saccharometer; the taste, amount of acidity, and any peculiarity of flavour are then noted.

The sediment should be shaken up with the few drops of beer remaining in the flask,* and a small drop examined in the usual way under the microscope. The chief points to be observed in the microscopic examination of these samples are :—

1st.—The condition of the original yeast (S. cerevisiæ).

2nd.—The amount of new yeast.

3rd.—The variety of secondary and wild forms present.

4th.—The presence or absence of Bacteria.

5th.—The forms·of Bacteria present.

The inter-comparison of such observations will lead anyone very quickly to form an opinion as to the *relative* keeping quality of the beers examined, and it is chiefly this relative value that is of service to the practical brewer, as it enables him to decide as to the order in which to send out his stock.

On Plate XII. may be seen examples of forced beer sediments, and it will be noted that the relationship of Saccharomyces to Bacterial forms is variable. From what we have already stated under the separate headings of the Bacteria and Saccharomycetes, it will be concluded that there must necessarily be a great diversity of appearance.

We have found it most convenient to group all forced samples into three classes, I., II., and III., and two sub-classes I. to II. and II. to III. All beers that taste sour when they come off the tray, and that have a high acidity (that is, above the normal but not actually sour), together with a deficiency of new yeast cells, and swarming with short and long Bacterial forms, we mark Class III., and should not consider it safe to keep such beers over six weeks from rack, unless treated with some antiseptic. Beers that have no objectionable peculiarity in flavour, and that when examined exhibit a fair

---

* Occasionally these sediments are very dark, almost black ; we have reason to believe this is due to a decomposition of the lead glass of which forcing flasks are often made, and the production of Sulphide of Lead, which is taken up by the cells.

amount of new normal yeast forms, yet containing a considerable number of Bacteria, we mark Class II., and should advise them to be put into the trade at an early date, anticipating difficulty with such beers if kept over two months. Sound-tasting beers exhibiting only normal secondary forms of yeast, or no new yeast, and free from Bacteria, or containing only a few Lactic or Bacillus forms in each field, we should mark Class I., and expect to stand well through the summer. Beers that on examination appear too good for Class II. and not good enough for Class I. we mark I. to II., and the same applies to the other sub-class.

The six examples given on Plate XII. are fairly typical of this arbitrary but convenient classification :—

Fig. 1.—A normal clean residue of S. Cerevisæ; the large size of some of the cells is probably a result of the forcing tray temperature = Class I.

Fig. 2.—A mixed growth of wild yeasts, chiefly S. Pastorianus and S. Ellipsoideus = Class I. to II.

Fig. 3.—An active growth of what is probably Pasteur's Caseous ferment (S. Coagulatus I.) in a form it is not infrequently met with in forced samples = Class I. to II.

Fig. 4.—Few ferment cells, and a vigorous growth of Bacillus subtilis, in the form most frequently seen in these deposits = Class II.

Fig. 5.—A few yeast cells and Bacteria and a con- siderable growth of Sarcina, probably Pediococcus acidi lactici = Class II.

Fig. 6.—Hardly any yeast growth, and swarming with rod and Bacillus forms, probably B. lactis and B. subtilis. If not above normal acidity = Class II. to III. If markedly acid = Class III.

The gas supply should be of a uniform and steady character, as one of the effects of the gas going out, and the tray consequently falling in temperature, is that mercury

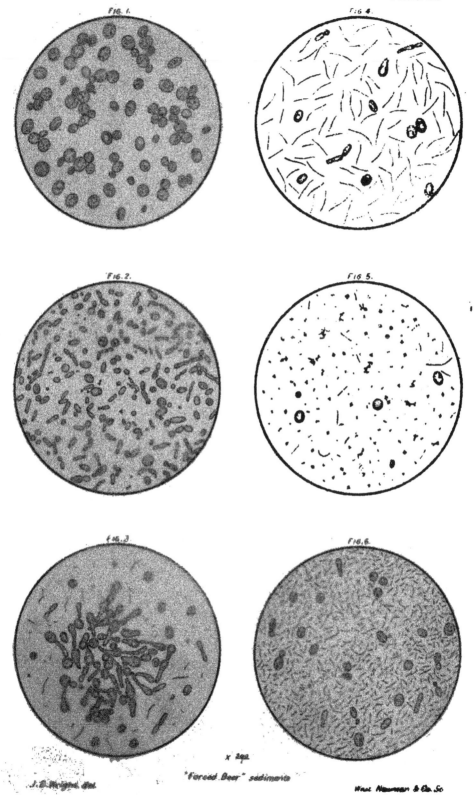

PLATE XII

FIG. 1.

FIG. 4.

FIG. 2.

FIG. 5.

FIG. 3.

FIG. 6.

× 200

"Forced Beer" sediments

J. E. Wright, del.

West, Newman & Co. Sc.

is driven back into the flasks, especially in the case of flat ales. In so far as the effect of the mercury itself is concerned, we have some reason to believe that it acts as an antiseptic, deterring the production and growth of Bacteria, and possibly of Saccharomyces ; but it is a different thing if air enters as well, as a clouding of the beer and film growth on the surface, together with acidity and occasionally ropiness, may ensue, and the experiment be altogether vitiated.

We have also found the tray a suitable apparatus for testing the tendency of malt extracts to become acid, by comparison of the amount of acidity formed in a given number of hours—from 72 to 120—with the normal acidity of the malt extract ; a free exposure to air and consequent infection being first permitted. We are of opinion that a useful factor in the determination of the quality of various samples of malt is thus obtained.

Another useful purpose to which the forcing tray may be placed is testing samples of water for purity by the so-called Heisch's test; in so far as the amount of Phosphates in a water is indicative of contamination. (See Chapter X.)

CHAPTER IX.

The Anatomy of the Barley-Corn.

IN the selection of a sample of barley, the Brewer or Maltster is guided by various features that are visible to the unaided senses, such as the general appearance of the corn, the nature of its skin and the state of the starchy portion ; but to comprehend the minute internal and external structure of the seed, we must apply a somewhat closer inspection by means of the microscope. Let us consider first in brief the more general attributes of the Barley-corn. We have a spindle-shaped body, somewhat more pointed at the germinal end, enveloped by a strong skin or husk (the *Paleæ*), which is fairly smooth and flat on the *dorsal* side, but considerably wrinkled and rounded on the other—the *ventral*. The dorsal side is traversed from end to end by five small ridges, caused by vascular bundles in the husk, and this latter is drawn down into a furrow, which extends along the corn on the ventral side. On looking closely into this furrow at the germ end, we discern in the perfect corn a small spike or bristle, which on being separated and placed under the microscope shows itself to be a bundle of fibres packed very closely together, with other small fibres or hairs standing out on all sides, presenting under a low power the appearance shown in Fig. 27 a. This is called the *Corn-bristle.* Now if, instead of breaking off this bristle, it is carefully dissected out under a hand-lens, from a steeped corn, or one that has been on the "floors" some days—damp corns being far more easily dissected than dry ones—it may be removed together with certain small processes attached to its base. To do this a sharp penknife or the sharpened needles previously mentioned may be employed, and the outer

Fig. 27.

J E W.

coating removed according to Fig. 28, the dotted lines being those of incision.

On examining this structure under a moderate power it presents the appearance indicated in Fig. 27 b—b, the portions so marked are known as *Lodicules.* They are oval transparent processes, somewhat resembling a hand with out-stretched fingers, the fingers being hairs or spines, similar to those on the bristle itself. The attachment between the Lodicules and the bristle is complete at c, where they unite with the inner Paleæ. Both these portions appear to be remnants of the flower of the Barley.

Now it is a property of minute tubes and bundles of fibres having small interstitial spaces, to absorb liquids freely, and this power is called Capillary attraction ; it may be seen to advantage if a glass tube be heated in a flame till it can be pulled out into a thread, and portions of this thread dipped at one end into a coloured fluid, such as red ink. The passage of the fluid up the capillary tube is plainly seen, the height to which it rises being determined by the diameter of the tube and nature of the liquid.

It has been argued* that the Corn-bristle and Lodicules together constitute an arrangement adapted for the capillary absorption of liquids, and it is on the face of it probable, that whether destined for this purpose or not, they are able to absorb water and carry it into the corn. It is however,

Fig. 28.

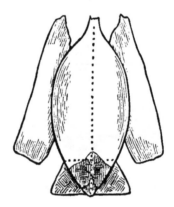

evident that absorption can go on without the intervention of the bristle, as corns from which it has become detached behave in the usual way on steeping ; the Lodicules, being protected by the husk, may nevertheless still convey water to the corn.

These views as to the function of the bristle and lodicules have been definitely refuted by recent researches, and experiments of our own indicate that the corn will take up water equally well at either end.

The husk of the Barley-corn, consisting of two leaf-like bodies, known as the *Palea*, may next be taken into

* " Die Anatomie des Gersten Kornes," Lorenz Enzinger.

Fig. 3.

Fig.

Outer layer of Palea × 300

Portion of Palea × 4

Fig 4

Fig 2

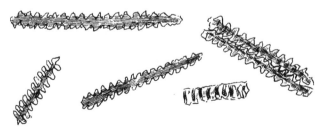

Disintegrated fibres of Paleæ × 300

C.G.MATTHEWS. DEL

BEMROSE & SONS LITH

consideration, and to do this effectively it is better to separate portions of and digest them for some days in warm dilute acid, dilute Caustic alkali, or Bromine water; they are then easily dissected and will be seen to consist of two distinct layers of cellulose fibres which when "in situ" lie in the direction of the length of the corn. This skin, magnified to a very moderate extent, is shown on Plate XIII., Fig. 1. The upper layer consists of toothed or corrugated fibres (Fig. 2), the corrugations dovetailing together as in Fig. 3, the round portions occurring at intervals being, as it were, pegs which connect the two layers, the lower of which is more distinctly fibrous, the fibres interlacing as at Fig. 4. Both layers are indicated in Fig. 1 in the position they naturally occupy. The whole forms a very strong and dense layer, yet possessing sufficient elasticity to meet the swell of the corn on steeping. The dorsal Palea, which in the unthrashed corn is continuous with the so-called "beard" or *awn*, just overlaps the ventral one at an equal distance on either side of the furrow.

The *Germ* which in sound corns becomes the young barley plant, lies beneath the inner transparent skin on the opposite side to the corn-bristle, and when the Palea is removed, appears as a small waxy yellow substance.

Beneath the Paleæ are two coats or skins, the one immediately underneath, called the *Pericarp*, is shown in Plate XIV., Fig. 1. It is a very fine integument, and exhibits when magnified a nearly transparent cellular structure, the cells having a general tendency to a rectangular form. The cells appear for the most part to be separated by minute spaces, and occupy a position with the longer axis of the cell in the same line as the longer axis of the corn. It is pretty certain that this second skin, by the nature of its structure and position, allows liquids to pass freely from end to end of the seed, and can take up water directly at either end where it is, so to speak, fractured by separation from the point of attachment to the ear and the "awn" respectively. The true covering of the seed or third skin, known as the *Testa*, is, like the Pericarp, a very fine layer of cellular matter ; the cells in this case having a decided tendency to a prismatic form, their longer axis being at right angles to a line drawn from end to end of the corn [Plate XIV., Fig. 2]. Here also the cells are separated by minute spaces which doubtless act as capillary tubes, and convey moisture around the inner seed. These two skins can be separated into various layers, but we consider it sufficient in this work to describe their main features only.

The inner Palea, together with the Pericarp and Testa, pass some way into the corn-furrow and fold there, but the layer of cells immediately underlying the testa passes considerably further into a central channel which extends the whole length of the corn, and is well seen in Plate XVI., Figs. 1 and 2, which represent transverse sections, Fig. 1 before germination has commenced, and Fig. 2 when it has proceeded some days. The channel thus formed completes the arrangement for the moistening of the interior of the seed, and for the circulation of liquids during the process of germination ; for being in direct communication with the absorbing tissues it can become filled with the water conveyed by them; and moisture being thus applied very completely to the starchy portions of the corn, all the inter-cellular spaces become filled up.

Having now considered the outer coatings, let us direct our attention to the inner seed, and refer to the longitudinal section of the barley-corn given in Plate XV. This section is supposed to be through the furrow.

A, represents the coatings generally, which have been already sufficiently described. The starchy portion or *Endosperm* B, is seen to be situated above the germinal

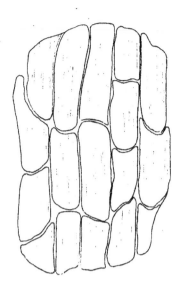

Fig. 1. Pericarp × 300

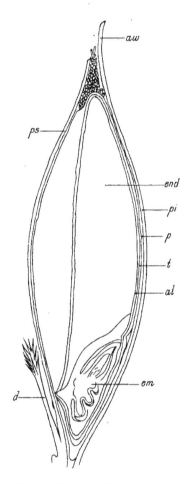

Fig 3. Diagram Section.
(after Holzner)

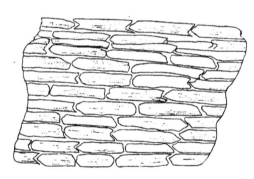

Fig 2. Testa × 300

ps Palea Superior   pi Palea inferior
al Aleurone layer   end Endosperm.
em Embryo   d   Basal bristle
p   Pericarp   aw Awn   t   Testa

Section of a Barley Corn in the plane of the long axis and through the furrow

(Reduced from C. Lintner)

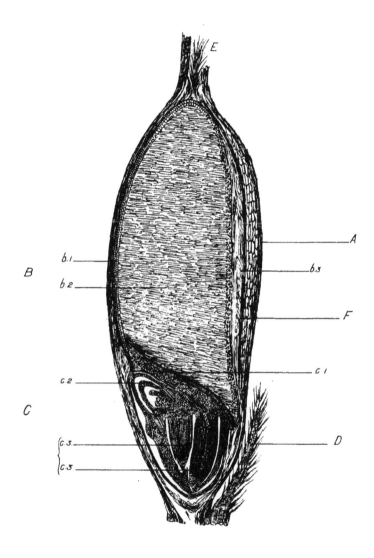

A Husk	B Endosperm	C Germ⊥ parts
D Corn bristle	b₁ Aleurone layer	c₁ Scutellum
E Pappus	b₂ Starchy matter	c₂ Acrospire
F Pigment string	b₃ Empty cells	c₃ Rootlets

Fig.1.

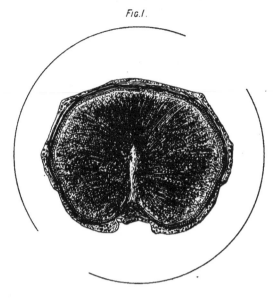

Fig.2.

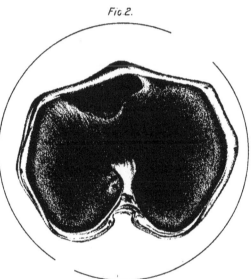

From a Photograph

Transverse sections of Barleycorn.

parts or *Germ* C. The Endosperm and the main bulk of the Germ are bounded by a peculiar layer known as the *Aleurone* cells, as well as by the Testa and Pericarp (see Plate XIV., Fig. 3). These cells, some of which are shown highly magnified in Fig. 29, contain finely granulated proteid or nitrogenous matter, and small spherules of fat or oil ; it is not clear what their immediate function is, but seeing that they are in contact with the starch cells of

Fig. 29.

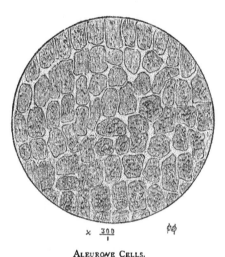

x 300/1

ALEUROИE CELLS.

the Endosperm and the great bulk of the Germ, they may take some active part in the transfer of food from the former to the latter.

The Endosperm itself is a mass of Starch cells, of which there are two kinds in the Barley-corn, large and small, intermingled with irregular and spherical particles of nitrogenous and mineral matter ; the whole contained in radial compartments of cellulose, and forming a store of food stuff to supply the germ until it is grown sufficiently to enable it to draw nourishment through its roots and leaves.

On disintegrating a portion of the Endosperm and examining microscopically, the larger starch granules are

easily distinguished, and can be rendered even more
distinct by staining with a little of the weak Iodine
solution (see Appendix), a drop or two being applied to
one side of the cover-glass, whilst a small piece of blotting
paper is held against the other side ; the Iodine solution
is thus carried across under the cover-glass. The starchy
portions assume a deep blue tint, and portions of matter
faintly coloured, or not coloured at all, are something other
than starch : owing to the diffusion of a small amount of
soluble starch in the corn, the non-starchy portions sometimes exhibit a shade of blue when thus treated.

The two kinds of Starch are shown in Plate XVII.,
Fig. 1. With oblique illumination obtained by suitable
openings in the diaphragm, or other means such as
staining with a solution of Chromic acid, concentric lines
are rendered visible on the starch granules, but they are
much more plainly seen on some other kinds of starch,
more especially Potato starch [Plate XVII., Fig. 2].
Examples of wheat, maize, and rice starches are given on
the same plate, and it will be seen that there is a considerable
variety in appearance.

The germ proper, which in the dried barley-corn forms
only a very small portion of the whole (about $\frac{1}{30}$), is
separated from the Endosperm by a sheath called the
Scutellum [Plate XV.], consisting of a dense *Epithelium*
of "palisade-like cells," upon which is usually found a layer
of compressed empty cells from which the starchy contents
have been dissolved.

Immediately underlying the upper end of the Scutellum
is the *Plumula* or *Acrospire* which, as germination proceeds,
gradually increases in size by cell multiplication, forces
its way beneath the Testa, and eventually emerges from the
upper end of the corn if its progress be not checked by
some such means as that adopted in malting ; at the same
time the embryo rootlets expand, separate, and descend

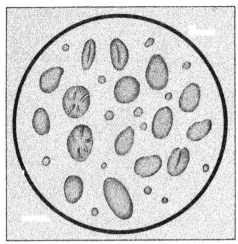

Barley Starch.

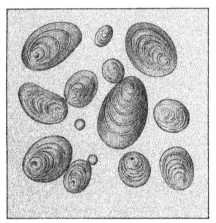

Potato Starch.

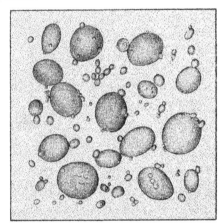

Wheat Starch

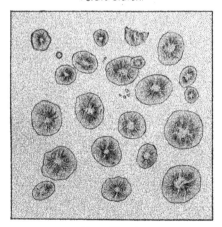

Maize Starch

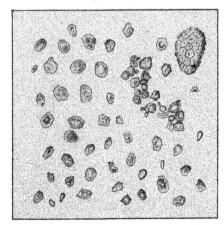

Rice Starch.

$$\frac{\times\ 350}{1}$$

J.E. WRIGHT. DEL.

BEMROSE & SONS Lt^

through the base of the corn. The number of rootlets varies according to the kind of barley, some barleys having only three, others as many as eight rootlets. As before stated the Endosperm contains the supply of food required by the germ, and the alteration of this supply from the almost insoluble non-diffusible state in which it originally exists, into a liquid that can be easily conveyed to the growing germ, is a point of very considerable chemical and biological interest. Horace T. Brown, in a most interesting paper on " A Grain of Barley,"* says that 40 % of the total reserve nitrogen originally present in the Endosperm, passed to the young growing plant in eleven days.

It has been noticed that during the growth of the Acrospire, the starch cells in its immediate vicinity are strongly influenced by some solvent, probably akin to Diastase, which dissolves away portions of them, creating an appearance called *pitting*. Two such pitted or eroded granules are shown in Plate XVII., Barley Starch. Of the soluble matters thus formed, a portion probably goes to nourish the germ, and it is not improbable that an active circulation is kept up by means of the various enveloping coats of the corn and the central channel, whereby a modification of the contents is continually going on so long as the corn is kept moist, the end result being the formation of fermentable sugars and diastase. Interesting, however, as these chemical changes may be, it would be plainly out of our scope to enter more fully into them, our object having been to describe the apparatus by which they are effected and which is beautifully adapted for the purpose.

In this sketch of the anatomy of the Barley-corn we have not attempted to go into detail as to the mode of development of the corn in the ear, nor to enter into considerations

* Transactions of the Burton Natural History and Archæological Society, 1889, p. 108.

that would appear to be purely of Botanical interest ; if, however, the reader would wish for further information he will find the subject exhaustively treated by Johannsen,* and lately by Holzner and Lermer,† who have practically worked out the minute anatomy of the grain. A concise epitome of this work is given in the paper by H. T. Brown, just referred to. We might add that it is not only useful but instructive for the student to make sections of the Barley-corn at different stages of its growth, and suggestions as to the preparation and permanent mounting of these sections will be found in the Appendix.

* Carlsberg Report, Vol. II., part 3.

† Beiträge zur Kentniss der Gerste.

CHAPTER X.

Hops, Sugar, and Water.

IN connection with hops there is comparatively little scope for the use of the microscope, seeing that the salient features of any given sample are, with the necessary experience, taken in by examination with eye and hand. We have in Chapter VI. alluded to the identification of moulds on the surface of hops, and we will now devote a little space to the consideration of one or two points in connection with the structure of the hop-cone, also known as the catkin or strobile, which, botanically considered, consists of a number of small bracts, with two ovaries at their base, each being accompanied by a rounded bractlet. Both bracts and bractlets enlarge greatly during the development of the ovary, and form, when fully grown, the membranous scales of the strobile.

The microscopic structure of the leafy portions of the Hop-cone has no direct interest for the Brewer apart from any appearance of mould, ravages of blight, or anything of a purely superficial character ; for the extractible substances that are of value in beers do not reside here, they are found in the " condition " or " lupulin." If the golden grains forming the latter be examined under a very moderate power, they are seen to consist of little vesicles or capsules, the form and structure of which is rendered quite plainly visible if their contents be exhausted by immersion in a small quantity of hot alcohol before placing on the slide. Fig. 30 shows the symmetrically-shaped capsule in its ordinary position at the base of the leaflet or bract. If a few of the capsules in their original state from new or recent hops, be broken on a glass slide by pressure on the cover-glass, the oily and resinous contents may be seen surrounding the broken capsule. This escaped matter includes the Hop-oil or Aroma, Hop-bitter proper, Resin, Fat, and astringent matter of the nature of Tannin. As hops 'age,' the contents of

Fig. 30.

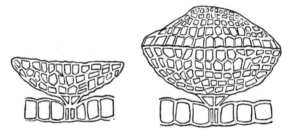

the capsules become gradually less oily and more highly coloured, till at length, in hops that are two or three years old, only hard dark-coloured matter is left, where formerly was a golden oily substance; the Brewer well knows the changes in the nature of the Hops accompanying such appearance. Interspersed amongst the capsules proper are found smaller vesicular bodies, consisting of four to eight cells grouped together, very much enlarged at the upper end; these vesicles are usually colourless. The nature of their contents has not, so far as we know, been definitely ascertained: in comparison with the large capsules, they would appear to be of quite secondary importance from a brewing point of view.

It is astonishing what a diverse collection of objects

may be removed from Hops by shaking them up several times with water, pouring the water off quickly, and by a fractional separation dividing the lighter objects from the heavier. Most hops thus treated show Bacteria, Crystals (probably Malate and Oxalate of Lime), Cells of Saccharomyces, Infusoria, and Protococcus. Many samples yield Mould-spores, and some few show these last in considerable profusion.

In Plate XVIII. are shown examples of most of these objects.

a. Ferment cells.

b. Bacteria.

c. Crystals.

d. Particles of Earth and Siliceous matter.

e. Spicules, probably part of the Hop plant.

f Mould-spores, probably of Ustilago and some species of Fusarium.

g. Probably Protococcus.

h. Probably Pollen cells.

i. Portions of Mould hyphæ.

By treating barley in a somewhat similar manner, and making a microscopic examination of the sediment, a variety of organisms, etc., is exhibited, which seems to be even greater than is obtained from hops. Plate XIX. shows the following objects so obtained :—

1. Starch cells.

2 and 4. Cells of protococcus.

3. Spicules, probably part of the corn.

5. Cells of Saccharomyces.

6, 7, 10, 13. Mould-spores (simple and compound), probably Ustilago Segetum and other species.

8, 9. Diatoms.

11 *a.* Pasteur's lactic ferment. *b.* Bact. aceti.

12 *a.* Bact. lactis. *b.* Bact. termo.

14. Pollen cells.

15 *a.* B. leptothrix.   *b.* Bacillus subtilis.

16. Compound spores of red mould.

The water first run off from steeping barley is a good source from whence the above and other organisms may be obtained ; the coarser particles should be separated by a short settling, and the liquid may then be left to deposit the finer particles for examination. No doubt most of the organisms found on Barley and Hops are discernible on other forms of vegetation freely exposed to the air during their growth.

It will from previous considerations be obvious that "dry-hopping" has its disadvantages as well as advantages, for many of the living cells, more especially those of Saccharomycetes, are able to, and often do set up in Beers a characteristic secondary fermentation, the so-called "hop-sickness," which in its early stages is sometimes attended by a very unpleasant smell. If the ale is inherently sound it will recover from this, but a faultily-brewed ale may not only support the alcoholic ferments introduced by hops—giving a persistent and awkward fret—but after these have had their sway, may have its decline hastened by the considerable addition of Bacteria beyond those it possibly contained at Racking. Any mould-spores introduced by dry hops would probably remain dormant till the Beer was drawn off, but might, under favourable conditions of growth, develop in the dregs and help to produce a mouldy cask.

SUGAR.

In cases where Brewing sugars do not dissolve to a clear solution in water, and give perhaps a well defined sediment, it is desirable to ascertain by the Microscope what this suspended or sedimentary matter may be ; more often than not, in the case of Glucoses or of Invert Sugar, it is

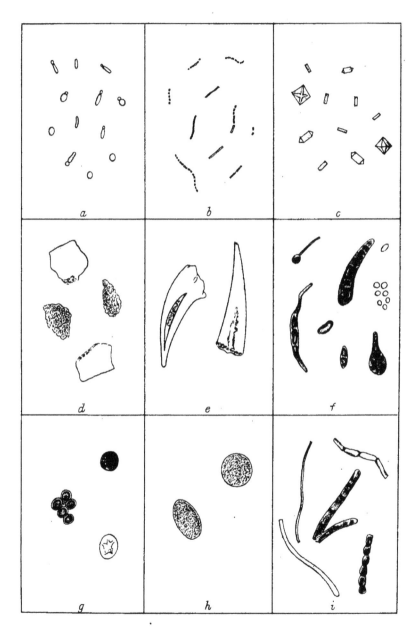

Organisms &c found in Hop dust.

× 300

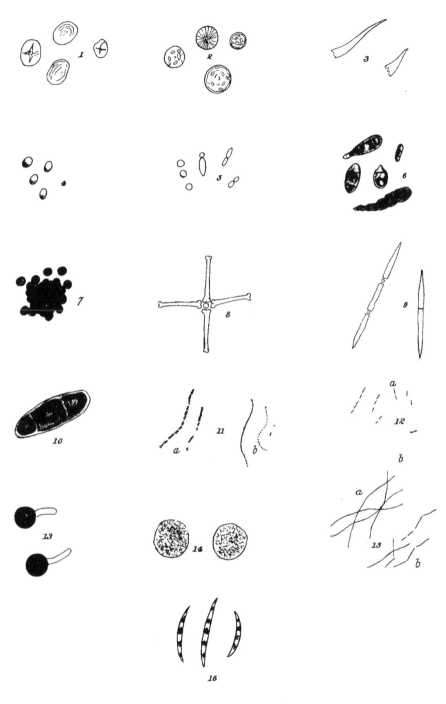

Organisms found in barley dust.

× 300

Sulphate of Lime left from the neutralizing ; the sulphate being soluble to a considerable extent, and perhaps crystallizing out in the concentrated syrup before solidification. There is nothing particularly objectionable in this, except that it does not indicate the most careful manufacture. Raw, unrefined, or partially refined Cane Sugars may show some diversity of extraneous objects, *e.g.*, Mould-spores, Saccharomyces, etc., and the insect Acarus sacchari is not unfrequently met with : in the latter case it is well that we have some confidence in the destructive action of the heat of the boiling copper, for it would be unpleasant to contemplate the possible survival of such organisms. Sugar solutions may be tested as to their power of supporting Bacteria, by dissolving one or two grams of the sugar in 250 c.c. of distilled water and putting in a clean corked or stoppered bottle on the forcing tray for a day or two. Sugars containing Phosphates or Phosphorus in organic combination develop Bacteria freely, and if a perceptible amount of phosphates be present, a Butyric fermentation will be set up. Such a state of things as this last, though not exactly indicating that the sugar is quite undesirable for all purposes, would nevertheless, we think, afford good ground for not using it in ales that were destined for " stock."

The microscope may be employed to ascertain the nature of the organisms that have developed in the solution treated as above.

The method of examining sugars just mentioned leads us to Heisch's test for potable waters, of which it is a modification. It is performed by taking about 250 c.c. of the water to be tested, and adding to it 1 to 1·5 grain of pure re-crystallized Cane Sugar. The bottle containing these is put on the Forcing Tank, and the appearance noted at different intervals during several days. Some waters remain quite clear, others become opalescent or milky, whilst those of the worst class go turbid and smell strongly of Butyric acid. The microscope will show the nature of the Bacteria present. The test, according to Prof. E. Frankland, indicates phosphates in the water, and this contention has been sustained by one of us in a series of experiments on a great many samples of water,* and as a rider to it, the fact has been established that Butyric fermentations occur in the waters containing most phosphates, other marked signs of contamination being at the same time afforded by chemical analysis.

Let us now speak of the suspended matters frequently contained by Brewing and other waters :—Besides mere earthy matter, we not unfrequently have to deal with a variety of organisms, including Bacteria, Moulds, and obviously living forms classed generally as Infusoria. The best treatment, of a water which appears likely to yield a sediment, is to shake up the containing vessel and pour a quantity of the water into a glass funnel holding about half-a-pint, closed at the narrow end by an inch or two of caoutchouc tubing, terminating in a small glass test-tube. After some hours settlement, most of the water may be poured off from the top ; the little tube is then quickly removed, and if necessary the supernatant water poured off from it separately, so as to leave the residue in a few drops of water only : this is shaken up and put on a slide. Plate XX. represents some of the objects we have thus obtained from a sample of town-supply water, which when freed from suspended matter was by no means impure.

$a =$ a Diatom.
$b =$ a Desmid, (?) in various stages of development.
$c =$ Monads (active).
$d =$ Earthy particles.
$e =$ portion of a Desmid.

* Jour. Soc. Chem. Ind., July 30th, 1887, Vol. VI., p. 495.

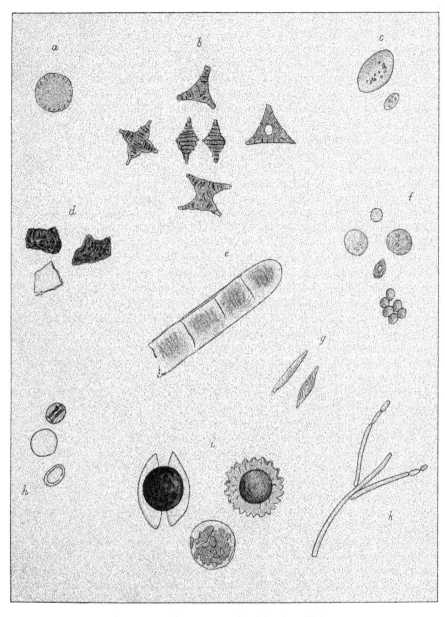

Organisms &c found in sample of Drinking Water

$\times \dfrac{300}{1}$

$f =$ Protococcus, in various forms.

$g =$ Diatoms.

$h =$ (?) Protococcus or Desmid.

$i =$ Desmids.

$k =$ piece of Mould growth.

The presence of these bodies may, and often does mean, that the containing reservoir, well, or tank is in an unclean state, and this of course ought to be remedied. The influx of sewage matter into a well by percolation is a rather distinct matter; here chemical analysis is the chief guide to the actual state of the water, but there are no doubt interesting and instructive results to be obtained by a Bacteriological investigation carried out by some process of cultivation in gelatine, such as the plate method we have already alluded to, and which has been lately described in detail by Dr. Percy Frankland.*

As an example of organisms in water we may quote Miquel, who found :—

35 germs per c.c. in rain-water caught as it fell.

62 ,, in river water from the Vanne.

1,400 ,, .. ,, Seine above Paris.

3,200 ,, ,, ,, ,, below Paris.

Dr. Percy Frankland† in November, 1885, found :—

1,866 germs per c.c. in Thames water at Hampton.

954 ,, in river Lea water at Chingford Mill.

Later he gives further figures representing the organisms present in 1 c.c. of the London water-supplies taken under different conditions; the results show more especially perhaps, how large a proportion of the organisms present (96% to 98%) is removed by the filtration carried out by the respective companies.

---

* Jour. Soc. Chem. Ind., Vol. IV., page 698.

† Ibid, page 706.

Koch holds that a good water never contains more than 150 individual mixed organisms in 1 c.c., and that the presence of any number much exceeding this is suspicious; 1,000 per c.c. rendering it unfit for drinking. According to other observers, this seems however to be a very arbitrary classification. Our own opinion is that this method of testing waters is at present only of the most general application, but that it may, as the knowledge of Bacteria advances, become of more importance.

Seeing that to all intents and purposes organisms contained in Brewing waters must either be killed by heat in the water itself or destroyed later in the Wort-copper, it is of more importance that the chemical constitution of the supply should be ascertained, than the fact that it contains so many Bacteria per cubic centimetre or per gallon. Doubtless a water containing Bacteria in plenty would, in many cases, prove on analysis to be contaminated; but it does not follow as a matter of course that it would be so.

Some waters, such as the Sulphur waters of certain springs, seem to afford a very suitable plasma for Bacteria and Microscopic fungi; varieties of Beggiatoa are for instance found in them, the growths being mainly long threads from 3 to 3.5 μ thick. The threads contain secreted Sulphur in grains, and by a process of decomposition give off Sulphuretted Hydrogen, causing the characteristic smell of certain waters.

Impure waters standing in wooden and even in metal tanks will, especially in warm weather, throw up a scum of living organisms; portions of this scum sink from time to time, eventually forming a layer of some thickness on the bottom.

A few words about the filtration of Brewery waters. Where any slight suspended matter is merely of an earthy character, filtration is hardly necessary, as such waters generally draw clear in time. If the suspended matter, however, consist of animalculæ, etc., it is more serious, and the state of the well and character of the supply should be investigated. If the supply *must* be used, an efficient filter is desirable. Amongst the best filtering media are Coke and Spongy Iron, whilst unglazed porcelain in the shape of the Chamberland filter seems to perform its office most perfectly.

Although the chemical constitution of a water may be somewhat changed by filtration, it is almost idle to suppose that it can turn a badly contaminated water into a pure and useful one ; at any rate, this amount of work is not yielded by any known filter on a practical scale. It is of the highest importance that a filter should not have its power overtaxed or be allowed to get clogged, for water after passage through filters in this condition, is generally rather more impure than before filtration.

CHAPTER XI.

Brewery Vessels, etc., etc.

HAVING, as we believe, given due consideration to the more direct applications of the Microscope to the brewing process, it remains for us to speak of cases where the instrument may be of service as accompanying or supplementing other modes of observation.

It has already been indicated that the air constitutes the immediate source of the organisms that may cause serious trouble in brewing, and so long as the present method of brewing obtains, aerial contamination may be regarded as a *constant*, and must be met by all proper precautions as to the employment of good materials and a well-considered method, thereby reducing the risk to a minimum.

The quantity of germs floating in the air of any given neighbourhood is as we have seen (page 109) very variable; depending mainly on actual contaminating influences, such as a dense population, free exposure of decomposing animal and vegetable matters, and prevailing dirt. Apart from these, certain atmospheric conditions have according to Miquel, corresponding effects, for instance :—

Prolonged rain purifies the air from bacteria, washing them into the soil; but they are re-dispersed when dust is again formed.

With a high barometer the number of germs in the air is proportionately greater, and less with a low barometer. Less also with a decrease in the amount of moisture. The proportion of ozone, and changes of temperature and of the direction of the wind, also affect the number. At sea the air is practically germ-free.

We have already had occasion to make passing mention of a series of experiments carried out by Hansen, to ascertain the nature of the organisms present in the air surrounding the Carlsberg Brewery, and in the buildings themselves.* For this purpose flasks of sterilized beer-wort were exposed to air infection, and the following organisms were identified, many of which are well known, and others have been already spoken of, but we think it desirable to reproduce the whole list.

SACCHAROMYCETES.

S. cerevisiæ.
S. ellipsoideus.
S. exiguus.
S. Pastorianus.
S. mycoderma.
S. apiculatus.
S. glutinis.

MOULDS.

Eurotium aspergillus glaucus.
Aspergillus fumigatus. ·
Penicillium glaucum.
    ,,   cladosporioides.
Mucor racemosus.
    ,,   stolonifer.
Botrytis cinerea.
Cladosporium herbarum.

* Meddelelser fra Carlsberg Laboratoriet, 1879 and 1882, vol. 1, part 4.

Dematium pullulans.
Oidium lactis.
A species of Dendrochium.
    ,,   ,,   Monilia.
    ,,   ,,   Arthrobotrys.
Indeterminate mycelium.

FORMS NOT CLASSIFIED, POSSIBLY MOULDS.

Cells like Saccharomyces cerevisiæ.
    ,,   Chalara.
Red cells resembling Saccharomycetes.
Small round cells of a "torula" form.

BACTERIA.

Bacillus subtilis.
    ,,   ruber.
Bacterium Kochii.
    ,,   pyriforme.
    ,,   Carlsbergense.
Mycoderma aceti.
    ,,   Pasteurianum.
Spirillum tenue.
Yellow bacillus.
Sarcina.
Micro-bacteria and Micrococcus.

Some parts of the Brewery showed more organisms than others. An elevated temperature favoured the production of organisms in the flasks exposed, and some of the organisms appeared even at 42° C. (107.6° F.), but such an elevated temperature more especially favoured Mycoderma Vini. Bacteria were particularly favoured by a temperature of 26° C. (78.8° F.)

A practical application of these researches has been made by washing and purifying the air entering the fermenting cellars of the Alt Carlsberg Brewery. The filtering and cleansing medium is brine, through which the air is allowed to pass, leaving behind the germs it contained.

In cases where fermenting worts were aerated by pumping machinery, we have seen filtration carried out by tying thicknesses of canvas over the inlet for air; or the air may be filtered through a kind of cushion containing cotton-wool not too tightly packed.

From foregoing matters it will be plainly recognized that from the time that worts on the " cooler " fall below a certain temperature, fully developed organisms finding their way into such worts, may retain their vitality unimpaired. The spores of some bacteria and probably of moulds, would resist even the highest temperature of " cooler " wort. When the opportunity arrives, these air-borne germs take effect, and such an opportunity is provided when the vitality of the yeast has been lowered by some of the various possible causes which we shall touch upon in the next chapter. A healthy, vigorous fermentation may be considered as precluding the development of disease organisms, and where the materials and process are good, and the pitching yeast clean—that is, free from bacteria and wild yeasts—air-borne germs are of little consequence, unless the air of the particular locality conveys an overwhelming number. As regards spores of moulds or bacteria surviving the boiling in copper, or introduced from the air, it may be said that malt actually having mould on it, is likely to be a cause of far more trouble than these, as it carries in itself the results of mould deterioration. In the same way a contaminated yeast carries its character stamped on it, and will, apart from the contained organisms, prove an inferior ferment.

Floors on which beer or wort is being constantly spilt are, if neglected, likely to get into a most offensive condition and may engender bacteria freely. Hot water is doubtless the best agent for cleansing wooden floors, and it is a good thing to occasionally follow up its use by mopping over with Bisulphite of lime. Old brick, cement, or tile floors that have become cracked or broken-up by age, may harbour all sorts of abominations in the way of bacteria, mould, etc. Renewal is about the only cure, but till this be effected bisulphite of lime or sulphurous acid may mitigate the evil.

The walls of fermenting and cleansing rooms should be kept in as good order as possible : no damp, mouldy, or clammy places should be allowed, but a clean surface free from dust and dirt provided. The surface of walls at the back of fermenting vessels, especially of "squares," sometimes gets into a deplorable state of dirt, or perhaps the word *filth* more correctly expresses the condition. This is usually as much a fault of construction as anything else, the places spoken of being almost inaccessible.

It is really astonishing how contaminated the air of racking rooms, tun-houses, etc., may become by neglect of thorough cleanliness, and good ventilation. For example:— our attention was on one occasion called to the state of a water-tank used for purposes of general supply in a tun-house or cleansing room. The water in this tank smelt badly although it emanated from a good source, and was not unfrequently renewed. On examination, a black sludge was found at the bottom, consisting of bacteria and yeast cells, most of the latter being stained black by contact with iron. The whole mass of sludge was developing sulphuretted hydrogen freely. The yeast and bacteria had doubtless mainly come from the air of the place, and had fallen into the tank, whose only covering consisted of a few loose boards.

It is of no little importance that the drains of a brewery should be in good working order and effectively "trapped."

Pipes for waste liquids from upper floors should, where convenient, discharge into properly constructed open gratings on the ground level, thus helping to avoid direct communication with the sewer; for sewer-gas is bad anywhere and must, if discharged into the brewery, help to convey organisms that have their proper place elsewhere.

A point worth noting in connection with the contamination of beers by foreign organisms is, that whilst the present method of "dry hopping" is pursued, it would be almost absurd to rigidly exclude bacteria, etc., during the manufacture of the beer, and subsequently introduce them with dry hops, by myriads, to the finished article: at the same time, all reasonable precautions are worthy of observance, to ensure freedom from excessive aerial contamination.

The most scrupulous cleanliness is, in our opinion, called for in the case of the surfaces of vessels—more especially the wooden ones—with which the worts and beer are in actual contact. In the first place, moist wooden surfaces seem to provide a not unsuitable *habitat* for bacteria and moulds, which may not only retain their vitality for a long time in the pores of the wood, but even multiply there; and thus a liquid contained in the vessels may, by its movements, detach and carry away active organisms from the surfaces under consideration. We have frequently had cause to examine shavings and portions of wood taken from old fermenting vessels, unions, union troughs, and tunning casks. By breaking these pieces up; soaking in water; pouring off the latter, and examining the sediment formed on standing a little while; a motley array of organisms is frequently exhibited, amongst which we have seen the following :—

Sarcina.
Bacillus ulna.

Ordinary rod and thread bacteria, probably B. lactis
and Bac. subtilis.

Moulds growing in the torula form.

Mould hyphæ and spores.

Even with the greatest care, wooden vessels must of
course deteriorate in time by wear and tear, and when the
wood becomes spongy it is almost impossible to secure
cleanliness.   Decay is greatly hastened, however, by
imperfect cleansing; for then bacteria, etc., have a better
chance of disintegrating the woody tissue.   The rational
course to pursue, is regular and thorough cleansing; replacing the vessels when really old, by new ones.

The foregoing remarks on brewery utensils apply quite
as strongly to cask plant, of which we have already spoken
in connection with moulds.

It is almost beyond question that of all disinfectants
bisulphite of lime and sulphurous acid—but preferably the
former—are the most effective and convenient for use in
connection with the cleansing of wooden vessels, for they
have a powerful action on both moulds and bacteria,
more especially perhaps on the latter.

The metal vessels of a brewery are, with the exception
of pipes, pretty easily cleansed; but pipes require special
attention and methods.   It often happens that a gelatinous
or leathery film is formed in pipes used for the convey-
ance of worts, etc., which film is not adequately removed
by brushing, and indeed can only be detached and cleared
away by strong, hot, caustic alkali.   Such films generally
contain bacteria in swarms, besides other organisms.

CHAPTER XII.

___

GENERAL REMARKS ON THE BREWING PROCESS.

___

ALTHOUGH we shall presently be travelling beyond the scope of the immediate application of the Microscope to the Brewing process, we do not think it will be altogether out of place if we offer some remarks on certain side issues that appear to us from their general interest to call for notice. We think it worth while at the same time to recapitulate some points already treated of.

As we have already shown in Chap. IV., all store yeasts may be regarded as mixtures, in which one type or species of Saccharomyces predominates according to the nature of the process; and where the results of this last are the most satisfactory, there is doubtless a greater persistency, and consequently a larger proportion of the ferment best adapted to the method of brewing pursued : in other words, the yeast is in unison with the character of the materials and process.

With a method unsuited to the persistence of a desirable species of yeast, there must be deterioration of the latter ; and the same state of things is arrived at, if a fresh pitching yeast be employed that is unsuitable to the process pursued ; as for example, trying to carry out a " Stone square" fermentation with Burton yeast, or pitching Burton worts with London yeast, and attempting to work them on the Burton system.

Deterioration of the store-yeast may be discovered by the change of character or inferiority of the ales produced, before it is apparent by the microscopical examination of the yeast itself. When traceable by the latter means, it may exhibit itself as follows :—(1) By the alteration in appearance of the cells of S. Cerevisiæ. (2) By the incursion of bacteria. (3) By the presence of wild yeast. The two former conditions are more easily distinguishable than the latter, which is sometimes only to be ascertained by fractional cultivation according to Hansen's method, or some modification of it ; as an example :—Some few years ago, one of us in conjunction with Mr. Wallis Evershed,* experimented on a reasonably pure-looking sample of Burton yeast, and by a process of separation, based on the degree of temperature at which different species of yeast were killed, the presence in the sample of S. Minor, S. Coagulatus No. 2, and spores of Mucor Racemosus, was demonstrated ; besides which, some very curious large pointed cells of yeast were obtained, which may, however, have been only modifications of S. Cerevisiæ, induced by the high temperatures to which the yeast was submitted to effect the differentiation. Now this being possible with a good average Burton yeast, it is clearly obvious that yeast from an irregular and faulty process would contain a large proportion of wild ferments ; and that it does so, is well-nigh a certainty in the majority of cases. We have met with pitching yeast that contained a large quantity of S. Pastorianus, and ascertained that the beers of the Brewery in which this particular yeast was produced, were liable to S. Pastorianus frets of a marked kind. Considering the variety of alcoholic

" Brewer's Guardian," vol. xiv., page 181.

ferments existing in nature, it is not surprising that store-yeasts are liable under suitable conditions, to become *mixtures* of them. Spontaneous fermentations of saccharine liquids exposed to air are nearly always carried out by a variety of ferments, though possibly one or more forms may preponderate according to the nature of the liquid ; for example, S. Apiculatus often appears and grows readily in the expressed ¡juice of the grape, but grows only with difficulty in beer-wort even when freely sown therein. A natural selection has doubtless taken place in the case of brewers' yeast which, from a general point of view, may be regarded as an educated or modified form from spontaneous or air-sown fermentation in the distant past ; all normal yeasts containing a predominating quantity of this naturally selected form. We may remark in dealing with this subject, that if fermentation of beer-wort be inaugurated by means ˙of barley-dust, a moderately pure and regular growth of yeast is generally obtained, which when separated from extraneous matters, dirt, etc., is indistinguishable from some ordinary pitching yeasts. It seems to us probable therefore, that S. Cerevisiæ is one of the ferment forms to be found on barley ; our experiments, although incomplete, point at least to this.

We have distinctly to insist upon the fact that brewery yeasts that appear pure and homogeneous may, and generally do, contain different species of Saccharomycetes of different degrees of persistency, according to the nature of the process. These species are often almost identical in appearance, and in one or two cases not very dissimilar in their fermentative action, and it is not until an abnormal percentage of one or other species appears that the presence of these foreign organisms is easily demonstrable ; though some time before this, the yeast may exhibit peculiarities and irregularities in its mode of action.

We have already alluded to the question of a possible

introduction into this country of pure yeast cultivation, and some points that arise in connection therewith. Whatever may be done in this direction, a considerable time must elapse before even the preliminaries of a practical issue are decided : in the meantime, the aim of every brewer should obviously be to study the conditions best suited to the production of a required type of yeast, and thus secure its continued reproduction in a state of relative purity. That this can be done, hardly admits of a doubt, for there are breweries in the United Kingdom producing an almost unvarying type of yeast, rendering any resort to outside " changes " practically unnecessary.

To treat of all the known or speculative causes of yeast deterioration would be going far beyond the province of this work ; at the same time it will occupy but little space to summarize the main causes, and the foregoing chapters will perhaps have indicated how far the Microscope is able to assist in identifying them. We may classify the causes as follows :—

1.—Those connected with materials :—Water, Malt, Hops, Sugar, etc.

2.—Those in connection with the process:—Temperatures employed. Periods of duration of certain operations. State of vessels, etc.

3.—The condition of the pitching yeast at any given time ; dependent mainly on the first two sets of causes.

The *well-known* causes of mishap to the brewer stated above in general terms, are :—

Impure steep water.
Indifferent or bad barley.
Unskilled malting.
Contaminated brewing water.
Unsuitable mashing temperatures.

Unsuitable fermenting temperatures, and inadequate control of the fermentations.

Markedly impure pitching yeast. •

A few remarks on some points arising from the above.

We have already taken cognizance of the fact that organisms—bacteria, moulds, etc.—occur on the surface of barley, and have also alluded to their presence in numbers in ordinary steep-water ; so that it will be readily understood, that an impure steep-water may not only bring these organisms much more quickly into active vitality, but may also furnish a supply of its own : a free growth of mould on the "floors" would be a very natural sequence, especially in mild or warm weather. In the case of impure brewing-waters, any contained organisms would in all probability be killed during the heating up for mashing : any deteriorating influence resulting from the constitution of the water being, of course, a purely chemical question.

In connection with fermentation, certain abnormal results are obtained from time to time, that seem to be traceable to "materials," rather than to the yeast itself. We may mention, more especially, "boiling" or "bladdery" fermentations, and "stenchy" fermentations. We have carefully examined yeast accompanying cases of this kind, without detecting anything distinctly unusual in its appearance. One or two cases of "bladdery" fermentation we have traced with certainty to slack malt, and one case of "stenchy" fermentation to sulphured hops ; but it is almost certain that other and more obscure influences may tend to produce "stench," this being caused by some modification of the fermentative action of the yeast not traceable by the microscope. It is hardly necessary to say that a "bladdery" fermentation does not produce a good crop of yeast, either as regards quality or quantity.

One of the chief conditions regulating the production of a yeast of uniform type is a reasonable uniformity in the character of the worts, and the mode of fermentation. The yeast must by careful selection, be kept in a certain state of equilibrium as regards chemical constituents; it must neither be impoverished by want of adequate nutrient matter, nor repleted by excess of the same, for instance : if a quick yeast like that of Burton be carried through consecutive worts of high gravity, a marked deterioration in fermentative vigour ensues, owing doubtless, to a repleted state of the ferment, which has become so rich in protoplasmic contents that saccharine solutions no longer exert their normal stimulating effect ; and it is quite possible that in addition, the cells are alcoholized or partially asphyxiated. Yeast thus deteriorated may be restored to activity by fermenting in a comparatively weak wort, and it therefore seems a fair argument that the surplus cell-constituents go to form new cells, without drawing so much on the cell-forming constituents of the wort. In contradistinction to the above, it not unfrequently happens that yeast becomes impoverished by consecutive growth in weak or average-gravity worts from a poor class of material, as also from the excessive use of sugar ; in the latter case an occasional all-malt brewing helps to restore the vigour of the yeast : in the former case, the same may be effected by putting the yeast through stronger worts, or worts from a better class of malt.

One of the most important influences on the well-being of yeast and its degree of activity, is that exercised by aeration or oxygenation. Yeast from quick fermentations will bear with advantage a very thorough exposure to air before being set in succeeding fermentations, and there can be no doubt that the worts should have a short but complete exposure to the air during cooling. These conditions are certainly found in the Burton system. With medium and slow fermentations caused by corresponding ferments, initial aeration does not seem to be quite so important ; the fermenting liquids, especially in the cases of the Scotch and Yorkshire Stone-square systems, receiving supplies of oxygen in detail by rousing in the first case, and rousing and pumping in the second ; the action of the yeast being also modified by low pitching heats and attemperation.

The yeasts produced by essentially different methods have an undoubted tendency to retain their particular habit, and consequently it is practically impossible to transform a fast yeast into a slow one, or *vice versâ*, in one operation.

We may now consider briefly the chief effects produced by yeast deterioration in beer itself, that is to say, effects more directly connected with the actual state of the pitching yeast. We have—

Sluggish fermentations.

Imperfect attenuation and cleansing.

Improper flavours.

Faulty behaviour in cask, *e.g.*, flatness, fret, persistent cloudiness, followed possibly by bacterial deterioration and finally, acidity.

The causes of many of these changes are well summed up in Pasteur's proposition ; " that every unhealthy change in the quality of beer coincides with the development of microscopic organisms which are alien to the pure ferment of beer." We have already described in some detail in previous chapters, the alcoholic ferments and bacteria associated with many of the changes for the worse, that beer undergoes ; in all such cases of change the microscope may be well applied as a first aid : at the same time it is evident that a final solution of the question must generally be sought in the chemistry and physics of the process, the microscope being, nevertheless, a valuable adjunct as a means of investigation.

M ICRO-PHOTOGRAPHY, also and perhaps more correctly termed Photo-micrography, is for those who have the inclination and leisure, not only a most interesting pursuit, but a far readier means of obtaining durable records of microscopic objects than can be secured by drawing.

The following apparatus is desirable :—a $\frac{1}{4}$-plate Camera, without a lens, or from which the lens can be readily detached, and the usual photographic accessories, including Dry-plates, Chemicals, Dark room, etc., or instead of this latter, the developing may be performed at night in a dark apartment, using a ruby lamp as source of light.

It is more convenient for the Camera to be placed vertically over the Microscope, and for this purpose two forms of stand are to be recommended. The first, which we ourselves use [Plate XXI.], has two upright iron rods carrying a movable wooden platform covered with cloth, having a circular hole about two inches in diameter in the centre, and side ledges which allow the camera to pass between, with little brass buckles to secure it firmly in position. A small distance is preserved between the camera face and platform by an inch-wide strip of thin cardboard or vulcanite close against each ledge ; this raises the Camera a trifle, and allows a thin sheet of vulcanite about three inches wide to be moved backwards and forwards freely between the Camera and platform. This

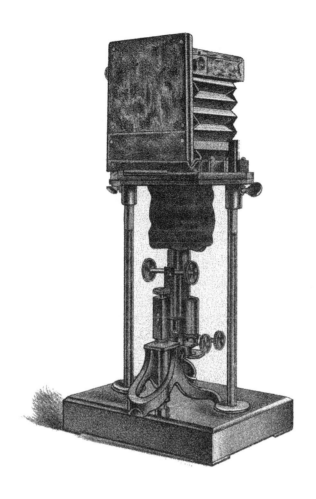

movable piece acts as a shutter, regulating the admission of light from the microscope tube to the ground-glass of the Camera.

The microscope, as will be seen by reference to the Plate, stands vertically under the platform, the end of the tube being enclosed in a black-velvet cylindrical bag, which at the upper end is fastened light-tight to the under side of the platform, and around its central opening. This bag allows various distances between the microscope tube and the platform, but does not interfere with the passage of light.

The other form of stand, which is simpler than ours, but must be strongly and firmly made to prevent shake, is virtually a large retort-stand. It consists of a platform having a strong metal rod fastened to it ; on this rod an arm can be adjusted at various heights by a clamping screw. The velvet bag may be used as before, and some arrangement is desirable for quickly fastening and unfastening the Camera.

Both these stands allow the observer to handle the adjustments of the microscope conveniently, and to look straight down on the image formed on the ground-glass screen of the Camera

Moderate sunlight or fairly strong artificial light is adequate for magnification up to about 200 diams., but from 250 and upwards, a very powerful paraffin or incandescent gas burner is at least necessary. Direct sunlight sometimes serves, but is as a rule very destructive of definition. An Oxy-hydrogen or Oxy-coal-gas lantern, though an expensive luxury, seems to be the most satisfactory source of illumination for high powers. A useful form of paraffin lamp was described in the November number of the " Society of Chemical Industry " for 1888 ; it is said to give very good results, and would we think, certainly do so with moderate magnifications.

The Microscope can be used with or without the eyepiece, according to the size of the disc obtained on the screen of the Camera, and the definition aecompanying either condition. The actual magnification can be determined, as in Chapter II., for any distance of Camera-screen from the top of the microscope tube, by photographing an ordinary micrometer scale, and comparing the value of the micrometer divisions with the lines depicted on the negative.

Space does not allow us to enter into details of ordinary photographic manipulation, they may well be gathered from a good handbook on photography, and better still by a practical exposition from some experienced photographer; for to see the operations skilfully performed is better than any amount of reading.

APPENDIX B.

On Preparing and Mounting Objects for the Microscope.

The objects encountered in the Brewing Process, permanent specimens of which may be desired, are :—Sections and dissected portions of the Barley-corn, Alcoholic ferments, Moulds, and Bacteria : any other objects being probably amenable to the treatment necessary for the forms mentioned.

If it be required to keep water-mounted objects—such as yeast, etc.—for some hours, and it is a case where glycerine is not advisable, we have found it a good plan to brush a little cedar-oil round the edge of the cover-glass, so as to seal in the water. Specimens may be preserved for some days and even weeks by this method. The oil can subsequently be easily removed with a little turpentine.

Sections of the Barley-corn are best made by embedding corns in the desired position in melted wax or paraffin-wax contained in an instrument called a Microtome, and then taking off shavings with a keen razor dipped in cold methylated spirit. The sections may be detached carefully from the surrounding wax, any remaining wax being dissolved away· by immersing them in turpentine. They may be mounted in Canada Balsam, but in this case the definition is not. good. Glycerine is a better medium, but it is very difficult to find a cement for the cover-glass edges. Gum dammar seems to stand fairly well. With Canada balsam the ordinary method of mounting would be as follows :—a scrupulously clean cover-glass and slide are taken, and on the latter a drop of Canada balsam is placed, which is judged sufficient to just extend to the margin of the cover-glass when this last is pressed down on it. The slide is gently warmed, and any air-bubbles are skimmed off the Balsam with a heated needle mounted in a wooden handle ; the section is taken out of the turpentine, the excess of the latter being removed by clean blotting paper, and is then introduced into the Balsam drop, any fresh air-bubbles being removed as before.

The cover-glass being warmed is now carefully put on, pressed down, and held by a small spring clip for some hours, until the Balsam has set. Any excess of Balsam may be removed by careful scraping, and after the lapse of a day or two, by careful cleaning round the edges of the cover-glass with an old silk-handkerchief dipped in turpentine.

In the case of glycerine mounting, the wax is mechanically removed from the sections, which are carefully immersed in a drop of slightly warmed glycerine ; air-bubbles being skimmed off, a cover-glass is pressed on, excess of glycerine wiped away, and the sealing medium applied with a brush on a Shadbolt turn-table. We have not tried it ourselves, but think it likely that sections might be preserved in a raised cell, or a glass cavity-cell, in a solution such as Goadby's (see Appendix C. III.), and finally sealed off on the turn-table with water-tight cement.

So far as we know there is no really satisfactory way of mounting yeast. Glycerine alters the appearance very markedly. Goadby's solution might answer, but would probably give opacity. A solid transparent medium yielded by a strong solution of white gelatine would enable specimens to be kept for a time. Possibly, drying off gradually in a drop of levulose solution might also serve.

Bacteria, after staining by some such method as that mentioned in Chap. VII., and drying off, can be moistened with a little turpentine or aniline oil, and a drop of Canada balsam laid on and treated in the manner described for mounting Barley sections. The Bacteria should previously have been diffused in a liquid which will not leave a residue of its own, otherwise the definition is not good.

Mould specimens, can like many others, be mounted dry in raised glass or wax cells ; the latter are made by cutting or punching out rings from a thin sheet of wax, paraffin-wax, or waxed cardboard, by means of sharp "cork borers" or punches dipped in methylated spirit : cardboard or paper rings dipped in melted wax may also be employed. The wax rings are dried and attached to the slide by slightly warming it to a temperature a little short of melting point of wax, and pressing the rings down gently with some flat surface. The growth may now be attached to the slide by the least touch of Canada balsam, and then the cover-glass slightly warmed and pressed down on the wax-ring. A protecting varnish can be laid on by means of the turn-table.

The appliances required for simple mounting are :—

Small quantities of Canada-balsam, Glycerine, Gum-dammar, Gold-size and Asphalt-varnish, solutions of Methyl-violet, Aniline magenta, and perhaps one or two other aniline dyes, Slides and medium-sized Cover-glasses, some wax and waxed-cardboard rings, a pair of small brass forceps, two or three needles in handles, one or two spring-clips for holding down cover-glasses, a spirit lamp or some other source of heat, a small copper or brass plate for drying off and warming, and a turn-table. Many of the accessories for mounting can, with a little ingenuity, be extemporized.

If further detail be desired, we should advise the student to consult some special work on the subject : a very useful little hand-book is " The Preparation and Mounting of Microscopic Objects," by Thomas Davies.

APPENDIX C.

VARIOUS PREPARATIONS FOR THE CULTIVATION AND PRESERVATION OF ORGANISMS.

I.—GELATINE FOR CULTIVATIONS. — In making this preparation, use preferably the transparent white thin-leaved gelatine.

First make a test solution to ascertain the consistency resulting ; using 1 oz. of gelatine to about 10 oz. of water in the following manner :—Break the plates of gelatine into small pieces, and soak for a few hours in cold water ; add the remainder of the water hot, and digest on a sand or water bath at about 160° F. till completely dissolved. The liquid filtered clear through felt bags can be collected in flasks, test-tubes, or in whatever vessels it may be required. The mouths of the vessels being closed with cotton wool, the gelatine may be completely sterilized by heating in an oven or water-bath to about 180° F. for about an hour or so. The original gelatine solution may be mixed with hopped or unhopped malt-wort, peptone, or other substances according to the purpose for which it is destined.

In connection with this subject we may refer to the following :—

"Jour. Soc. Chem. Ind.," 1885, p. 698, Percy Frankland.
Ibid, 1886, p. 114, ,, ,,
Ibid, 1887, p. 113, G. H. Morris.
"Brewers' Guardian," June 12th, 1886, C. G. Matthews.
Ibid, July 26th, 1887, J. G. Nasmyth.

II.—Nutrient Solutions.—

Raulin's Fluid.

	Parts by weight.
Water	... 1,500·0
Sugar Candy	70·0
Tartaric Acid	4·0
Nitrate of Ammonia	4·0
Phosphate of Ammonia	0·6
Carbonate of Potassium	0·6
Carbonate of Magnesia	0·4
Sulphate of Ammonia	0·25
Sulphate of Zinc ...	0·07
Sulphate of Iron ...	0·07
Silicate of Potassium	0·07

Pasteur's Solution.

150 cc. of a 10 % solution of pure Sugar Candy, 0·5 gramme of Yeast Ash obtained in a cupel furnace, 0·2 grm. of Ammonio-dextro-tartrate, and 0·2 grm. of Ammonic Sulphate.

Pasteur's fluid, with Yeast Ash replaced by Chemicals.*

Water	8,576 parts.
Cane Sugar ...	... 1,500 ,,
Ammonium Tartrate	100
Potassium Phosphate	2
Calcium Phosphate	2
Magnesium Sulphate	2 ,,

As Ammonium Salts rather inadequately replace Organic Nitrogen, the Ammonium Tartrate may well be replaced by a smaller quantity of Pepsin.

III.—Preservative Solutions for Vegetable Tissues.

Glycerine and Gum.

Pure Gum Arabic	1 oz.
Glycerine	1 .
Distilled water ...	1 ,,
Arsenious acid ...	1½ grains.

Dissolve the Arsenious acid in the cold water, then the gum, finally add the Glycerine, and mix without bubbles.

Goadby's Fluid.

Rock Salt	1 oz.
Alum	½ ,,
Corrosive sublimate	1 grain.

Dissolve in 1 pint of boiling water and filter. We found this last to answer well for keeping some specimens of germinating Barley.

IV.—Cement for Ring-cells or Finishing.

India rubber	½ drachm.
Asphaltum	4 oz.
Mineral Naphtha	10 ,,

Dissolve the India-rubber in the naphtha, then add the asphaltum. If necessary, heat must be employed, but only with great precaution.

* "Elementary Biology." Huxley and Martin, 1875.

13

APPENDIX D.

REAGENTS OR TESTING LIQUIDS.

Iodine Solution.

(For Starch granules, Bacteria, etc.)

To $\frac{1}{2}$ oz. of water and $\frac{1}{2}$ oz. of alcohol add a few crystals of Potassic Iodide and a few grains of Iodine. A portion of this solution may be diluted with water, till the colour is that of a full golden sherry.

Methyl-Violet.

(For staining yeast or Bacteria.)

Dissolve a small piece of violet copying-pencil lead, or the dye itself—which may be easily procured—in distilled water ; dilute till quite transparent.

Solutions of Bismark brown, Eosene, Aniline magenta, and Picric acid are all easily made, if required. Where the substance is not readily soluble in water, a little alcohol may be added as well.

Weak Ammonia.

(For clearing away resin.)

A few drops of strong Ammonia per 1 oz. water. A very weak Caustic Soda or Caustic Potash solution may be employed for the same purpose : the Ammonia solution, however, keeps better.

APPENDIX E.

THE STORAGE OF PITCHING YEAST

Seems to us a matter well deserving of notice. Cool vessels—such as slate—in a clean, cool, dust-free position, are desirable at all times, but more especially so in summer and autumn, when, as we know, there is the greatest risk of aerial contamination. Attemperators are a useful adjunct to yeast storage vessels, if a low temperature cannot otherwise be secured.

Where yeast has deteriorated to such an extent that some cleansing operation is necessary before it is used for pitching, it is obvious that an immediate change is the most desirable thing ; still a brewer may be so situated that he is obliged to go on with his own yeast, and in such a case the following observations may prove of some service :—

In the first place, when yeast is left to itself and is slightly " on the work," there is a tendency for the bacteria to come to the surface, owing possibly to their affinity for oxygen ; consequently, if a vessel of store yeast that has been standing some time is skimmed, a proportion of the disease organisms may be removed. A further purification may be effected by mixing the yeast with about ten times its volume of cold water in a somewhat shallow vessel. After standing an hour or two, amorphous matter and dead cells are deposited, the remaining yeast and water being run off from this layer. A further settling of 6 to 8 hours in a cold place—once more running off the liquid—provides a moderately clean yeast, which may be re-invigorated for use by mixing with a little sweet wort of about 1030 Sp. Gr. at a low temperature, some hours before pitching. Where yeast is very impure, Salicylic acid dissolved in a little Carbonate of Soda or Borax solution, may be employed with advantage in the proportion of about 1 oz. per Barrel of wash water.

APPENDIX F.

Foreign Pressed Yeast

Varies so markedly in its appearance under the Microscope, that only the most general description can be given. The cells are of varying size and shape, generally well vacuoled and nucleated, with sometimes a considerable tendency to elliptic forms. The impurities are usually, Bacteria (often B. lactis) and admixed starch, generally of Potato. The power of pressed yeast as a panification ferment can only be determined by actual experiment, being a function connected with the temperatures at which the yeast has been grown, and more especially connected with the particular species of Saccharomycetes, some species being naturally good panification ferments, whilst it is only with difficulty that the power can be developed in others.

We may here remark that samples of pressed yeast, wrapped in sterilized blotting paper, may be kept for a considerable length of time in a state of comparative purity.

Hansen uses alcohol and a 10 % solution of Cane Sugar for preserving yeast.

Samples of yeast may be kept for many months or even years, by careful air-drying and intimate admixture with plaster of Paris ; or by a suitable mixture of whole meal and wheat or potato starch with the liquid yeast, prior to pressing and air-drying. In each case the success of the operation depends upon the dried samples being kept absolutely free from moisture, it is therefore advisable to cover the corks of the bottles in which they are kept with paraffin-wax.

181

GLOSSARY;

OR,

EXPLANATION OF SCIENTIFIC AND TECHNICAL TERMS.

Note.—*Syn.* = *Synonymous with.*

A.

ABERRATION, an unequal deviation of the rays of light.

,, CHROMATIC, a fault in lenses which causes them to split up white light into its component colours, giving images with coloured edges.

,, SPHERICAL, a fault in lenses and mirrors which causes them to concentrate light to more than one focus, giving images with indistinct or blurred outlines.

ACARUS SACCHARI, a small animal allied to the cheesemite, found in common raw sugars.

ACHROMATIC, applied to a lens free from chromatic aberration.

ACROSPIRE, the bud of a germinating barley-corn. Syn. Plumule.

AEROBIAN, term applied to ferment forms induced by growth with free access of air.

AECIDIUM BERBERIS, a mould occurring on the Berberry plant, derived from Puccinia graminis or "rust" of corn.

ALBUMEN, botanically speaking, the contents of the barley-corn and other seeds. Applied by chemists to white of egg, and allied substances found in many living bodies.

ALEURONE, a peculiar layer of cells surrounding and partly constituting the mealy portion of the barley-corn and other seeds.

ALTERNATION OF GENERATION, the occurrence at definite intervals of a distinctly different form of growth in the consecutive genera-tions of living things, *e.g.*, moulds.

AMPLIFICATION, the enlargement of a magnified image.

ANTISEPTIC, a substance that prevents or delays putrefaction.

APLANATIC, applied to a lens free from spherical aberration.

ARTHROBOTRYS, a mould having a clustered appearance.

ASCOSPORES, spores formed in a sac-like cell called an ascus.

ASCUS, a cell or sac in which spores are formed.

ASPERGILLUS, a group of moulds, the spores of which are readily dispersible.

ASPERGILLUS FUMIGATUS, a mould having a smoky appearance.

,, GLAUCUS, a mould having a bluish green appearance.

,, NIGER, a black mould.

AWN, the beard or spike of a barley-corn.

B.

BACILLUS, a name given to short rod forms of Bacteria.

BACILLUS AMYLOBACTER, the starch-producing bacterium. Syn. Clostridium butyricum.

,, LEPTOTHRIX, a long, thin, hair-like bacterium.

,, RUBER, a bacterium having a red appearance in cultivations.

,, SUBTILIS, a thin rod-bacterium. Syn. the hay bacillus.

,, ULNA, a thick jointed rod-bacterium.

BACTERIUM–A, a general term applied to the Schizomycetes or fission-fungi.

BACTERIACEÆ, term applied by Zopf in his classification of the Bacteria, to a group including a variety of forms, and amongst them short rods.

BACTERIUM ACETI, a bacterium producing acetic acid.

,, BUTYRICUM, a bacterium producing butyric acid. Syn. Bacillus amylobacter; Clostridium butyricum.

,, FUSIFORME, a bacterium having a spindle shape.

,, PYRIFORME, a bacterium having a pear shape.

,, XYLINUM, a bacterium producing cellulose.

BEGGIATOA, a microscopic organism found in certain waters.

BRACT and BRACTLET, a small leaf more or less changed in form, from which a flower or flowers proceed.

C.

CAMERA LUCIDA, a light-reflecting apparatus, applied to the microscope for drawing objects.

CAPILLARY, hair-like.

CAPSULES, in botany applied to a seed-case, sometimes applied to the resin glands or lupulin of the Hop.

CARBOHYDRATE, a chemical term for substances such as Sugar, Starch, etc., which contain Carbon, and Oxygen and Hydrogen in the proportions in which they exist in water.

CASEOUS, a term applied to certain yeasts which have a tendency to fall from a liquid as a curdy precipitate.

CATKIN, in botany, a form of flower like that of the willow and hop. Syn. Strobile.

CELLULOSE, a carbohydrate forming the main constituent of all vegetable cells. *Pith* is nearly pure cellulose.

CHALARA MYCODERMA, a mould forming a loose white or grey film on liquids.

CHLOROPHYLL, the green colouring matter of plants.

CHROMATIC ABERRATION, See Aberration.

CHROMOGENOUS, applied to the bacteria having the power of producing colour.

CILIUM–A, minute hair-like filaments which act as the motile organs of bacteria, etc.

CLADOSPORIUM HERBARUM, a mould found on plants.

CLADOTHRIX DICHOTOMA, a thread-like bacterium, exhibiting the peculiarity known as false-branching.

CLADOTRICHEÆ, term applied by Zopf in his classification of bacteria, to the forms exhibiting false-branching.

CLEANSING CASKS, technical term for vessels in which beer is cleansed of yeast

CLOSTRIDIUM BUTYRICUM, a short thread bacterium. Syn. Bacterium butyricum ; Bacillus amylobacter.

COCCACEÆ, term applied by Zopf in his classification of the bacteria, to the forms which chiefly appear as small rounded cells, cocci or micrococci.

COCCUS (a berry), term applied to the small rounded form of many bacteria.

CONCAVE (hollowed), applied to lenses and mirrors having a hollowed surface.

CONDENSER, an apparatus for concentrating light on an object under microscopic examination.

    ,,   BULL'S-EYE, formed of a glass like that in a bull's-eye lantern.

CONDITION, term applied to the yellow resin-glands of the hop ; also to beers in a state fit for consumption.

CONIFERÆ, an order of plants, like the fir and pine, which bear their seeds in cones.

CONVEX, a term applied to curved lenses and mirrors, the curve falling away or downwards from the centre.

COOLER, technical term for a flat shallow vessel in which beer-wort is cooled.

CRENOTHRIX KUHNIANA, a bacterium occurring in wells and drain-pipes.

CRYPTOGAMIA, one of the two great divisions of the vegetable kingdom, consisting of the flowerless plants.

CULTIVATION, the growth of any particular organism which has been sown in a prepared medium.

D.

DEMATIUM PULLULANS, a mould frequently occurring in ripe fruit.

DENDROCHIUM, a white arborescent mould.

DESMID, a microscopic organism found in water.

DESMOBACTERIA, term applied by Cohn in his classification of the bacteria, to the thread-like forms.

DIASTASE, a soluble ferment produced in germinating seeds, which is capable of converting starch into sugar and gum.

DIASTATIC, the property of diastase.

DIATOM, a microscopic fresh-water Alga (seaweed), having a cell wall or " valve" formed largely of Silica (sand), with regular geometric markings.

DIPLOCOCCUS, two rounded bacteria more or less closely joined together.

DORSAL, applied to the outward side of a seed *in situ.*

E

ENDOSPERM, the internal matter of seeds—such as the barley-corn—upon which the young plant feeds during its early growth.

ENDOSPORE, a spore formed in an ascus or in the body of a ferment cell.

EPICARP, term applied to the outer layer of the pericarp or seed case.

EPITHELIUM, the fine membranous lining of the internal organs of all living things.

ERYSIPHE TUCKERI, a mould on cereals and vines. Syn. Oidium Vini.

EUROTIUM ASPERGILLUS GLAUCUS, an alternation form of Asp. Glaucus.

    ,,   ORYZÆ, a mould, the spores of which are found in Kôji, the Japanese ferment.

F.

FIELD, term applied by microscopists to that part of the slide under observation, seen at any given time through the instrument.

FLAGELLUM, a whip-like appendage possessed by many microscopic organisms, enabling them to move freely in liquids. Syn. Cilium.

FOCUS, the point to which rays of light or heat are concentrated by a lens or mirror.

FORCING TRAY, apparatus used for keeping vessels at a constant degree of warmth.

FORCING FLASK, a glass vessel for testing beer or other liquids on the forcing tray.

FUNGUS, a class of non-flowering, leafless plants (Thallophytes).

FUSARIUM HORDEI, the red mould of barley, having spindle-shaped or crescent spores.

G.

GLANDS, applied, in botany, to special cells containing particular substances, such as oil, resin, etc.

H.

HÆMATIMETER, a glass slide, so ruled and fitted, that microscopic objects placed on it may be measured or counted; originally used for counting blood corpuscles.

HYPHA, the tube-like, stem-forming cells of moulds or fungi, often forming a web or net-work.

I.

INFUSORIUM–A, microscopic organisms found in water and other liquids.

K.

KÔJI, macerated rice containing fungus spores, used by the Japanese as a ferment in making Saké, and bread; also in the manufacture of "Soy."

L.

LAGER, term applied to store beer brewed on the "low" or bottom" fermentation system.

LEPTOTRICHEÆ, term used by Zopf, in his classification of Bacteria, to the long thread and spiral forms.

LEUCONOSTOC MESENTERIODES, a bacterium which occurs in white gelatinous masses in the expressed juice of beet-root.

LODICULE, a dried-up part of the flower of grasses, etc., which remains attached to the seed or grain.

LUPULIN, term applied to the resin-glands of the hop flower.

M.

METRE, the standard of length of the French metric system : approximately 39·37 inches.

MESOCARP, the middle layer of the Pericarp or seed-case.

MICROBACTERIA, term used by Cohn, in his classification of Bacteria, for oblong cells which at times occur in gelatinous groups.

MICROBE, general term for microscopic organisms of the nature of bacteria.

MICROMETER, an instrument applied to the microscope for measuring small objects or spaces.

MICRON, term now used to express the thousandth of a millimetre. Syn. Micromillimetre of Botanists and Biologists.

MICROTOME, an instrument used for cutting extremely fine slices of objects for examination under the microscope.

MILLIMETRE, the thousandth part of a metre.

MOLECULAR, belonging to, or consisting of the groups of atoms, of which all substances are believed to consist.

MONAD, a microscopic animalcule found in water.

MONILIA CANDIDA, a white film-forming mould.

MONOCULAR, a microscope having one tube and eyepiece.

MUCEDINES, term applied to the moulds generally, but more correctly to a small division of them.

MUCORINI, term applied by Nägeli to the moulds; by De Bary to one group of them only.

MUCOR MUCEDO, a mould of very common occurrence.

,, RACEMOSUS, a mould very similar to M. Mucedo, forming a clustering mycelium in liquids.

,, STOLONIFER, a mould bearing black sporangia; the hyphæ tend to re-enter the nutrient stratum.

MYCELIUM, that part of a fungus or mould—formed usually of interlaced hyphæ—which corresponds with the root of other plants.

MYCODERMA ACETI, a film or "mother" forming organism.

,, VINI, an ærobian ferment; forms what is usually called "mother" of wine and beer.

MYCOPROTEIN, an albuminous or nitrogenous substance forming an essential part of living cells, more especially of fungi.

N.

NUCLEUS, term applied to granules found in the vacuoles of the yeast cell; and to somewhat similar granules, in living cells generally, which originate new cells.

O.

OIDIUM LACTIS, a mould occurring on stale milk.

,,   LUPULI, a mould occurring on spent hops.

,,   VINI, a mould found on the vine, and not unfrequently appearing in wine. Syn. Erysiphe Tuckeri.

OVARY, that part of a flower in which the seeds are formed.

P.

PALEA-Æ, small leaf-like bodies attached to many flowers; in the case of barley, forming the outer coating of the corn: they form the "chaff" of other cereals.

PAPPUS, the downy hairs at the summit of the ovary in certain plants, including barley.

PASTEURIZED, sterilized by heat as recommended by Pasteur.

PATHOGENIC, that which causes disease; applied to the bacteria associated with certain definite diseases.

PEDIOCOCCUS ACIDI LACTICI, a small bacterium which produces lactic acid.

,,   ALBUS, a small bacterium giving white cultivations.

,,   CEREVISIÆ, a small bacterium.

PELLICLE, a thin film; term applied by Hansen to the surface growth of certain yeasts.

PENICILLIUM GLAUCUM, a mould of a bluish-green colour.

,,   CLADOSPORIOIDES, a mould occurring on the shoots of plants.

PERICARP, that part of a fruit covering the seeds; one of the thin coatings of the barley-corn.

PERITHECIUM, a flask or cup-shaped receptacle, containing the spore-sacs or asci of a mould or fungus.

PITCHING YEAST, technical term for yeast used for starting a fermentation.

PLANE, a perfectly level surface, which may be at any inclination.

PLASMA, material giving rise to living matter.

PLEOMORPHY, existence of an organism in more than one form.

PLUMULE, or stem-bud, that part of the germ of a seed which ultimately becomes the stem of the young plant. Syn. Acrospire

POLLEN, the fertilizing powder on the stamens or male organs of flowers.

POLYMORPHISM, the existence of an organism in many forms. Syn. Alternation of generation.

PROTEID, similar in composition to protein.

PROTEIN, a substance containing Nitrogen and various other constituents found in living things.

PROTOCOCCUS, a single-celled fresh-water alga (seaweed): the green dust on tree stems, old wood, etc., and the green slime in water.

PROTOPLASM, an albuminous or nitrogenous substance forming an essential part of all living cells.

PSEUDOSPORES, false spores.

PUCCINIA GRAMINIS, a mould found on wheat and other cereals. Syn. "rust" of wheat.

R.

RAY, a single line of light or heat.

REAGENT, chemical term for any liquid or solid substance used to detect the presence of other substances.

REFLECTION, the turning back of a ray of light or heat from a polished or bright surface.

REFRACTION, the bending of a ray of light on passing into a medium of different density.

ROUND, technical term for a vessel in which fermentation takes place.

S.

SACCHAROMYCETES, the ferments which split up sugar and form alcohol.

SACCHAROMYCES APICULATUS, a ferment having a pointed form.

,, CEREVISIÆ, the usual ferment of beer.

COAGULATUS, a ferment having a curdy appearance when suspended in liquids. Syn. Caseous yeast.

CONGLOMERATUS, a ferment in which the cells are clubbed together in a curious manner.

ELLIPSOIDEUS or ELLIPTICUS, a ferment having elliptical cells.

EXIGUUS, a ferment of small size.

MARXIANUS, a ferment described by Marx.

MEMBRANÆFACIENS, a ferment forming a film or membrane.

MINOR, a ferment of a small rounded form.

PASTORIANUS, a ferment first described in detail by Pasteur.

GLUTINIS, a ferment giving rose-coloured slimy spots on potato, etc.

SAKÉ, a fermented liquid made in Japan.

SARCINA AURANTIACA, a small bacterium producing a golden yellow appearance.

,, CANDIDA, a small bacterium giving snow-white cultivations.

,, FLAVA, a small bacterium producing a yellow colour.

SARCINA GLUTINIS, a small bacterium.

,,   LITORALIS, a small bacterium occurring in sea-water.

,,   MAXIMA, the largest bacterium of the Sarcina family.

SCHIZOMYCETES, the Bacteria or fission-fungi.

SCUTELLUM, the membrane dividing the starchy part (endosperm) of the barley-corn from the germ.

SEPTUM, a partition or division.

SPHÆROBACTERIA, term applied by Cohn in his classification of bacteria, to round cells which at times occur in gelatinous groups.

SPHÆROTHECUM CASTAGNEI, a mould occurring on the hop-plant. Syn. Hop mildew.

SPICULE, a minute slender point.

SPIRILLUM TENUE, a thin spiral bacterium found in decomposing liquids.

,,   VOLUTANS, a spiral revolving bacterium found in decomposing liquids.

SPIROBACTERIA, term applied by Cohn, in his classification of bacteria, to the spiral forms.

SPORANGIUM, a receptacle containing spores.

SPORES, the seeds of flowerless plants, and of the lowest forms of animal life.

SPORULATION, the act of forming spores.

SQUARES, technical term for vessels in which fermentation takes place.

STERILIZE, to render free from living organisms of any kind.

STILLIONS, technical term for vessels in which beer is cleansed of yeast.

STROBILE, a form of flower such as that of the hop. Syn. Catkin.

T.

TESTA, the true skin of a seed.

TETRACOCCUS, a group of four cocci or minute bead-like cells.

THALLOPHYTES, a group of leafless, non-flowering plants, including algæ, fungi, and lichens.

TUNS, technical term for vessels in which beer is cleansed; also applied to Brewing vessels generally.

U.

UNIONS, trade term for vessels in which beer is cleansed of yeast.

USTILAGO CARBO, a black mould, the " smut " or " brand " of corn.

,,   SEGETUM, a black mould, the " smut " of corn ; found especially on cereals.

V.

Vacuole, cavity in the protoplasm of most cells filled with cell sap.

Ventral, applied to the inward side of a seed *in situ.*

Vesicle, a little bladder ; any small membranous cavity in animals or vegetables.

Vibrio, a term applied to the short bacteria which have a rapid movement.

Y.

Yeast-Flasks. Syn. forcing-flasks : a misnomer.

Z.

Zoogloea, term applied to a gelatinous colony of bacteria.

Zoospore, a motile spore of certain moulds.

Zygospore, a large spore produced in some moulds by a kind of sexual fructification.

INDEX

BEMROSE AND SONS, PRINTERS, DERBY AND 23, OLD BAILEY, LONDON.

Lightning Source UK Ltd.
Milton Keynes UK
UKOW06f0753051216
289221UK00017B/804/P